A Concise Course in Continuum Mechanics

A Concise Course in

Continuum Mechanics

Jiashi Yang

University of Nebraska-Lincoln, USA

World Scientific

NEW JERSEY • LONDON • SINGAPORE • BEIJING • SHANGHAI • HONG KONG • TAIPEI • CHENNAI • TOKYO

Published by

World Scientific Publishing Co. Pte. Ltd.
5 Toh Tuck Link, Singapore 596224
USA office: 27 Warren Street, Suite 401-402, Hackensack, NJ 07601
UK office: 57 Shelton Street, Covent Garden, London WC2H 9HE

Library of Congress Control Number: 2026001692

British Library Cataloguing-in-Publication Data
A catalogue record for this book is available from the British Library.

A CONCISE COURSE IN CONTINUUM MECHANICS

ISBN 978-981-98-2388-8 (hardcover)
ISBN 978-981-98-2389-5 (ebook for institutions)
ISBN 978-981-98-2390-1 (ebook for individuals)

For any available supplementary material, please visit
https://www.worldscientific.com/worldscibooks/10.1142/14611#t=suppl

Desk Editors: Kannan Krishnan/Amanda Yun

Typeset by Stallion Press
Email: enquiries@stallionpress.com

Preface

Continuum mechanics is usually offered as a graduate course in an engineering school to graduate students from different engineering disciplines. During the past two to three decades, the backgrounds of the students taking the continuum mechanics course changed significantly. Although most students have taken the standard undergraduate mechanics and mathematics courses, many do not have any experience in and have no plan to take graduate-level courses on the theory of elasticity, fluid mechanics, and methods of mathematical physics or partial differential equations. Overall, for various reasons, current graduate students taking the continuum mechanics course are less prepared in mechanics and mathematics compared to a couple of decades ago. To these students, existing continuum mechanics books often prove to be too challenging because of the breadth and depth of the topics covered and the heavy use of advanced mathematical techniques. Therefore, there is a need for an introductory and concise textbook on continuum mechanics suitable for a one-semester course to students with limited or no knowledge of graduate-level mechanics courses and advanced mathematical skills.

This book intends to present a concise and modern treatment of continuum mechanics suitable for a one-semester graduate course. A unified theoretical framework of thermomechanical phenomena in a continuum is constructed. It is broad enough to include nonlinear elastic solids and fluids with thermal and dissipative effects. Advanced and relatively involved special topics as well as complicated mathematical tools are avoided or minimized. Some relatively simple special topics are included as exercises.

The book is organized in a simple and straightforward manner as follows. Chapter 1 presents the basics of the indicial notation and Cartesian tensors, which are the common language of modern continuum mechanics. Chapter 2 discusses continuum kinematics, i.e., the description of deformation and motion. In Chapter 3, the relevant physical laws (balance laws) are postulated in integral forms. Their differential forms and other implications are systematically developed. Chapters 1 through 3 are valid for both solids and fluids. Chapter 4 is on constitutive relations in which specific material behaviors come into play. Constitutive relations for nonlinear thermoviscoelastic solids and nonlinear Stokesian viscous fluids are established and are reduced to various special cases. Chapters 5 and 6 treat elastic solids and fluids, respectively. Governing equations including the balance laws and constitutive equations are grouped together for various materials. Some simple solutions to a few initial-boundary-value problems are obtained for elastic solids or fluids.

Books on continuum mechanics are many. Only those that are used directly in the present book are listed as references. No attempt was made to review the literature. Because of the multiple variables, parameters and fields involved, some symbols may have different meanings in different places of the book. For convenience, a list of symbols is given in Appendix 1. Some basic equations in cylindrical and spherical coordinates are collected in Appendices 2 and 3. Appendix 4 gathers a few vector identities used in various places in the book.

About the Author

Jiashi Yang received his bachelor's and master's degrees from Tsinghua University, China, in 1982 and 1985, and his Ph.D. from Princeton University, USA, in 1994. Then he did postdoctoral research in the US at the University of Missouri–Rolla and Rensselaer Polytechnic Institute, and worked as an engineer with Motorola, Inc. He was a professor at the Department of Mechanical and Materials Engineering of the University of Nebraska-Lincoln, USA from 1997 to 2026. Since 2026, he has been employed by the School of Mechanical Engineering and Mechanics of Ningbo University, China. His research focuses on the mechanics of electromagnetoelastic structures and devices. His previous books include *Mechanics of Piezoelectric Structures*, *Mechanics of Functional Materials*, *Mechanics of Ferromagnetoelastic Materials and Structures* and *Theory of Electromagnetoelasticity* with World Scientific, as well as *An Introduction to the Theory of Piezoelectricity*, *Analysis of Piezoelectric Semiconductor Structures* and *A Concise Course in Elasticity* with Springer.

Contents

Chapter 1

Index Notation and Cartesian Tensors

This chapter is a brief treatment of the indicial notation and Cartesian tensors needed for the rest of the book. In Sections 1.1–1.5, we rewrite knowledge from calculus and linear algebra on vectors, matrices, determinants, scalar and vector fields using the index notation and summation convention. Cartesian tensors begin with Section 1.6.

1.1 Index Notation

Consider the Cartesian coordinate system in Fig. 1.1.

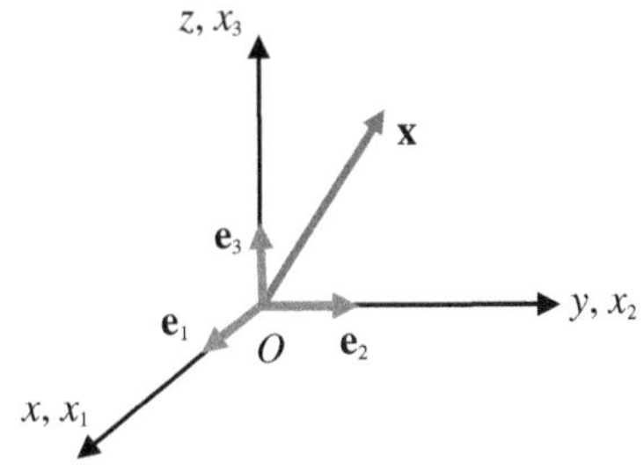

Fig. 1.1. A Cartesian coordinate system and the position vector of a point.

With the index notation, the components of the position vector $\mathbf{x} = x\mathbf{i} + y\mathbf{j} + z\mathbf{k}$ of a point are written as x_i with i assuming 1, 2 or 3, i.e.,

$$(x, y, z) = (x_1, x_2, x_3). \tag{1.1}$$

The index i in x_i is called a free index. x_k represents the same position vector as x_i. Similarly, we write the basis vectors as

$$(\mathbf{i}, \mathbf{j}, \mathbf{k}) = (\mathbf{e}_1, \mathbf{e}_2, \mathbf{e}_3) = (\mathbf{i}_1, \mathbf{i}_2, \mathbf{i}_3). \tag{1.2}$$

Then

$$\mathbf{x} = x_1\mathbf{e}_1 + x_2\mathbf{e}_2 + x_3\mathbf{e}_3 = \sum_{i=1}^{3} x_i\mathbf{e}_i = x_i\mathbf{e}_i, \tag{1.3}$$

where we have introduced the summation convention that repeated indices are summed from 1 to 3. Obviously,

$$\mathbf{x} = x_i\mathbf{e}_i = x_k\mathbf{e}_k. \tag{1.4}$$

Repeated indices are called dummy indices. They appear in pairs. Different pairs of dummy indices in the same term should be in different symbols.

We define the following nine numbers (Kronecker delta) by

$$\delta_{kl} = \begin{cases} 1, & k = l \\ 0, & k \neq l \end{cases} = \delta_{lk}. \tag{1.5}$$

It can be verified that

$$\mathbf{e}_k \cdot \mathbf{e}_l = \delta_{kl}. \tag{1.6}$$

We also have

$$\delta_{kk} = \delta_{jj} = \delta_{11} + \delta_{22} + \delta_{33} = 3. \tag{1.7}$$

As an example,

$$\delta_{1j}x_j = \delta_{11}x_1 + \delta_{12}x_2 + \delta_{13}x_3 = x_1. \tag{1.8}$$

Similarly,

$$\delta_{2j}x_j = x_2, \quad \delta_{3j}x_j = x_3. \tag{1.9}$$

In summary,

$$\delta_{ij}x_j = x_i. \tag{1.10}$$

Hence, the multiplication of x_j by δ_{ij} effectively changes the index of x_j from j to i.

As another example,

$$\delta_{1j}\delta_{js} = \delta_{11}\delta_{1s} + \delta_{12}\delta_{2s} + \delta_{13}\delta_{3s} = \delta_{1s}. \tag{1.11}$$

Similarly,

$$\delta_{2j}\delta_{js} = \delta_{2s}, \quad \delta_{3j}\delta_{js} = \delta_{3s}. \tag{1.12}$$

In summary,

$$\delta_{ij}\delta_{js} = \delta_{is}. \tag{1.13}$$

We also defined the permutation symbol with 27 components by

$$\begin{aligned} \varepsilon_{ijk} &= \begin{cases} 1 & i,j,k = 1,2,3; \quad 2,3,1; \quad 3,1,2, \\ -1 & i,j,k = 3,2,1; \quad 2,1,3; \quad 1,3,2, \\ 0 & \quad \text{otherwise,} \end{cases} \\ &= \mathbf{e}_i \cdot (\mathbf{e}_j \times \mathbf{e}_k). \end{aligned} \tag{1.14}$$

Since the three components of $\mathbf{e}_i$ are $(\delta_{i1}, \delta_{i2}, \delta_{i3})$ or $\mathbf{e}_i = \delta_{i1}\mathbf{e}_1 + \delta_{i2}\mathbf{e}_2 + \delta_{i3}\mathbf{e}_3$, Eq. (1.14) can be written as

$$\varepsilon_{ijk} = \mathbf{e}_i \cdot (\mathbf{e}_j \times \mathbf{e}_k) = \begin{vmatrix} \delta_{1i} & \delta_{2i} & \delta_{3i} \\ \delta_{1j} & \delta_{2j} & \delta_{3j} \\ \delta_{1k} & \delta_{2k} & \delta_{3k} \end{vmatrix}. \tag{1.15}$$

Now that ε_{ijk} can be written as a mixed triple product of three vectors or a determinant, we have, from the properties of determinants,

$$\varepsilon_{ijk} = \varepsilon_{jki} = \varepsilon_{kij} = -\varepsilon_{kji} = -\varepsilon_{jik} = -\varepsilon_{ikj}, \tag{1.16}$$

which are also evident from the definition of the permutation symbol. For $\mathbf{e}_i$, we have

$$\mathbf{e}_2 \times \mathbf{e}_3 = \mathbf{e}_1, \quad \mathbf{e}_3 \times \mathbf{e}_1 = \mathbf{e}_2, \quad \mathbf{e}_1 \times \mathbf{e}_2 = \mathbf{e}_3, \tag{1.17}$$

which can be written as

$$\mathbf{e}_j \times \mathbf{e}_k = \varepsilon_{ijk}\mathbf{e}_i. \tag{1.18}$$

For a verification of Eq. (1.18), we write it as

$$\mathbf{e}_j \times \mathbf{e}_k = \varepsilon_{ijk}\mathbf{e}_i = \sum_{i=1}^{3} \varepsilon_{ijk}\mathbf{e}_i = \varepsilon_{1jk}\mathbf{e}_1 + \varepsilon_{2jk}\mathbf{e}_2 + \varepsilon_{3jk}\mathbf{e}_3. \quad (1.19)$$

In the special case when $j = 2$ and $k = 3$, Eq. (1.19) reduces to

$$\mathbf{e}_2 \times \mathbf{e}_3 = \varepsilon_{123}\mathbf{e}_1 + \varepsilon_{223}\mathbf{e}_2 + \varepsilon_{323}\mathbf{e}_3 = 1\mathbf{e}_1 + 0\mathbf{e}_2 + 0\mathbf{e}_3 = \mathbf{e}_1. \quad (1.20)$$

Similarly, it can be verified that Eq. (1.19) also leads to

$$\mathbf{e}_3 \times \mathbf{e}_1 = \mathbf{e}_2, \quad \mathbf{e}_1 \times \mathbf{e}_2 = \mathbf{e}_3. \quad (1.21)$$

The following relationship exists between the Kronecker delta and the permutation symbol (ε–δ identity) [1,2]:

$$\begin{vmatrix} \delta_{ir} & \delta_{is} & \delta_{it} \\ \delta_{jr} & \delta_{js} & \delta_{jt} \\ \delta_{kr} & \delta_{ks} & \delta_{kt} \end{vmatrix} = \varepsilon_{ijk}\varepsilon_{rst}. \quad (1.22)$$

It can be verified directly. For example, when $(i, j, k) = (r, s, t) = (1, 2, 3)$,

$$\begin{vmatrix} \delta_{11} & \delta_{12} & \delta_{13} \\ \delta_{21} & \delta_{22} & \delta_{23} \\ \delta_{31} & \delta_{32} & \delta_{33} \end{vmatrix} = \begin{vmatrix} 1 & 0 & 0 \\ 0 & 1 & 0 \\ 0 & 0 & 1 \end{vmatrix} = 1. \quad (1.23)$$

Or, from Eq. (1.15), we can write

$$\begin{vmatrix} \delta_{1r} & \delta_{1s} & \delta_{1t} \\ \delta_{2r} & \delta_{2s} & \delta_{2t} \\ \delta_{3r} & \delta_{3s} & \delta_{3t} \end{vmatrix} = \varepsilon_{rst}. \quad (1.24)$$

Then the multiplication of Eqs. (1.15) and (1.24) leads to

$$\begin{vmatrix} \delta_{ir} & \delta_{is} & \delta_{it} \\ \delta_{jr} & \delta_{js} & \delta_{jt} \\ \delta_{kr} & \delta_{ks} & \delta_{kt} \end{vmatrix} = \varepsilon_{ijk}\varepsilon_{rst}. \quad (1.25)$$

The proof of Eq. (1.25) is left as an exercise at the end of the chapter (Problem 1.7). If we expand the determinant in Eq. (1.25) by the first row, we have

$$\begin{aligned}\varepsilon_{ijk}\varepsilon_{rst} &= \delta_{ir}\begin{vmatrix}\delta_{js} & \delta_{jt}\\ \delta_{ks} & \delta_{kt}\end{vmatrix} - \delta_{is}\begin{vmatrix}\delta_{jr} & \delta_{jt}\\ \delta_{kr} & \delta_{kt}\end{vmatrix} + \delta_{it}\begin{vmatrix}\delta_{jr} & \delta_{js}\\ \delta_{kr} & \delta_{ks}\end{vmatrix}\\ &= \delta_{ir}\begin{vmatrix}\delta_{js} & \delta_{jt}\\ \delta_{ks} & \delta_{kt}\end{vmatrix} - \begin{vmatrix}\delta_{is}\delta_{jr} & \delta_{jt}\\ \delta_{is}\delta_{kr} & \delta_{kt}\end{vmatrix} + \begin{vmatrix}\delta_{it}\delta_{jr} & \delta_{js}\\ \delta_{it}\delta_{kr} & \delta_{ks}\end{vmatrix}. \qquad (1.26)\end{aligned}$$

Setting $r = i$ in Eq. (1.26), we obtain

$$\begin{aligned}\varepsilon_{ijk}\varepsilon_{ist} &= 3\begin{vmatrix}\delta_{js} & \delta_{jt}\\ \delta_{ks} & \delta_{kt}\end{vmatrix} - \begin{vmatrix}\delta_{js} & \delta_{jt}\\ \delta_{ks} & \delta_{kt}\end{vmatrix} + \begin{vmatrix}\delta_{jt} & \delta_{js}\\ \delta_{kt} & \delta_{ks}\end{vmatrix}\\ &= \begin{vmatrix}\delta_{js} & \delta_{jt}\\ \delta_{ks} & \delta_{kt}\end{vmatrix} = \delta_{js}\delta_{kt} - \delta_{jt}\delta_{ks}, \qquad (1.27)\end{aligned}$$

which will be used often.

1.2 Vector Algebra

In a rectangular coordinate system, a vector **a** can be written in component form as

$$\mathbf{a} = a_1\mathbf{e}_1 + a_2\mathbf{e}_2 + a_3\mathbf{e}_3 = \sum_{i=1}^{3} a_i\mathbf{e}_i = \sum_{j=1}^{3} a_j\mathbf{e}_j = a_i\mathbf{e}_i = a_j\mathbf{e}_j, \qquad (1.28)$$

where, in the index notation, the components are a_i. Vector addition and the multiplication of a vector by a scalar are written as

$$\begin{aligned}(\mathbf{a}+\mathbf{b})_i &= a_i + b_i,\\ (\lambda\mathbf{a})_i &= \lambda a_i.\end{aligned} \qquad (1.29)$$

The dot or scalar product of two vectors takes the following form:

$$\begin{aligned}\mathbf{a}\cdot\mathbf{b} &= a_i\mathbf{e}_i\cdot b_j\mathbf{e}_j = a_ib_j\mathbf{e}_i\cdot\mathbf{e}_j = a_ib_j\delta_{ij}\\ &= a_ib_i = a_jb_j = a_1b_1 + a_2b_2 + a_3b_3. \qquad (1.30)\end{aligned}$$

The cross or vector product of **a** and **b** is defined as

$$\mathbf{a}\times\mathbf{b} = (a_2b_3 - a_3b_2)\mathbf{e}_1 + (a_3b_1 - a_1b_3)\mathbf{e}_2 + (a_1b_2 - a_2b_1)\mathbf{e}_3 \qquad (1.31)$$

which can be formally written as

$$\mathbf{a} \times \mathbf{b} = \begin{vmatrix} \mathbf{e}_1 & \mathbf{e}_2 & \mathbf{e}_3 \\ a_1 & a_2 & a_3 \\ b_1 & b_2 & b_3 \end{vmatrix}. \tag{1.32}$$

In the index notation, we have

$$\begin{aligned} \mathbf{a} \times \mathbf{b} &= a_i\mathbf{e}_i \times b_j\mathbf{e}_j = a_ib_j\mathbf{e}_i \times \mathbf{e}_j = a_ib_j\varepsilon_{ijk}\mathbf{e}_k = \varepsilon_{ijk}a_ib_j\mathbf{e}_k \\ &= \varepsilon_{123}a_1b_2\mathbf{e}_3 + \varepsilon_{231}a_2b_3\mathbf{e}_1 + \varepsilon_{312}a_3b_1\mathbf{e}_2 \\ &\quad + \varepsilon_{321}a_3b_2\mathbf{e}_1 + \varepsilon_{213}a_2b_1\mathbf{e}_3 + \varepsilon_{132}a_1b_3\mathbf{e}_2 \\ &= a_1b_2\mathbf{e}_3 + a_2b_3\mathbf{e}_1 + a_3b_1\mathbf{e}_2 - a_3b_2\mathbf{e}_1 - a_2b_1\mathbf{e}_3 - a_1b_3\mathbf{e}_2 \\ &= (a_2b_3 - a_3b_2)\mathbf{e}_1 + (a_3b_1 - a_1b_3)\mathbf{e}_2 + (a_1b_2 - a_2b_1)\mathbf{e}_3, \end{aligned} \tag{1.33}$$

or

$$\begin{aligned} &\mathbf{a} \times \mathbf{b} = \varepsilon_{ijk}a_ib_j\mathbf{e}_k, \\ &(\mathbf{a} \times \mathbf{b})_k = \varepsilon_{ijk}a_ib_j = \varepsilon_{kij}a_ib_j. \end{aligned} \tag{1.34}$$

As an example, we prove the following vector identity [3] using the index notation:

$$\mathbf{a} \times (\mathbf{b} \times \mathbf{c}) = (\mathbf{a} \cdot \mathbf{c})\mathbf{b} - (\mathbf{a} \cdot \mathbf{b})\mathbf{c}. \tag{1.35}$$

Denote

$$\begin{aligned} &\mathbf{d} = \mathbf{b} \times \mathbf{c}, \\ &d_k = (\mathbf{b} \times \mathbf{c})_k = \varepsilon_{kmn}b_mc_n. \end{aligned} \tag{1.36}$$

Then

$$\begin{aligned} [\mathbf{a} \times (\mathbf{b} \times \mathbf{c})]_i &= \varepsilon_{ijk}a_jd_k = \varepsilon_{ijk}a_j\varepsilon_{kmn}b_mc_n \\ &= \varepsilon_{kij}\varepsilon_{kmn}a_jb_mc_n = (\delta_{im}\delta_{jn} - \delta_{in}\delta_{jm})\, a_jb_mc_n \\ &= a_jb_ic_j - a_jb_jc_i = a_jc_jb_i - a_jb_jc_i \\ &= [(\mathbf{a} \cdot \mathbf{c})\mathbf{b}]_i - [(\mathbf{a} \cdot \mathbf{b})\mathbf{c}]_i. \end{aligned} \tag{1.37}$$

1.3 Matrices

We consider 3×3 matrices only. Let

$$\begin{bmatrix} A_{11} & A_{12} & A_{13} \\ A_{21} & A_{22} & A_{23} \\ A_{31} & A_{32} & A_{33} \end{bmatrix} = [A_{kl}] = \mathbf{A}. \tag{1.38}$$

The elements of $\mathbf{A}$ are written as A_{ij}. In the index notation, the trace of $\mathbf{A}$ can be written as

$$\mathrm{tr}(\mathbf{A}) = A_{11} + A_{22} + A_{33} = \sum_{i=1}^{3} A_{ii} = A_{ii} = A_{kk}. \tag{1.39}$$

The transpose of $\mathbf{A}$ is given by

$$\mathbf{A}^T = \begin{bmatrix} A_{11} & A_{21} & A_{31} \\ A_{12} & A_{22} & A_{32} \\ A_{13} & A_{23} & A_{33} \end{bmatrix}, \tag{1.40}$$

or

$$A_{ij}^T = A_{ji}. \tag{1.41}$$

A matrix is symmetric if it satisfies

$$\mathbf{A} = \mathbf{A}^T, \quad A_{ij} = A_{ji}. \tag{1.42}$$

A matrix is antisymmetric if

$$\mathbf{A} = -\mathbf{A}^T, \quad A_{ij} = -A_{ji}. \tag{1.43}$$

Matrix addition and the multiplication of a matrix by a scalar take the following form:

$$\begin{aligned} (\mathbf{A} + \mathbf{B})_{ij} &= A_{ij} + B_{ij}, \\ (\lambda \mathbf{A})_{ij} &= \lambda A_{ij}. \end{aligned} \tag{1.44}$$

A matrix can be broken into the sum of two matrices, one is symmatric and the other is antisymmetric as follows:

$$\begin{aligned} &A_{ij} = B_{ij} + C_{ij}, \\ &B_{ij} = \frac{A_{ij} + A_{ji}}{2} = B_{ji}, \quad C_{ij} = \frac{A_{ij} - A_{ji}}{2} = -C_{ji}. \end{aligned} \tag{1.45}$$

The multiplication of two matrices can be written as

$$\begin{aligned} \mathbf{C} &= \mathbf{AB}, \\ C_{ij} &= \sum_{k=1}^{3} A_{ik}B_{kj} = A_{ik}B_{kj}. \end{aligned} \tag{1.46}$$

The identity or unit matrix is

$$\boldsymbol{\delta} = \mathbf{1} = \begin{bmatrix} 1 & 0 & 0 \\ 0 & 1 & 0 \\ 0 & 0 & 1 \end{bmatrix}, \tag{1.47}$$

and

$$\boldsymbol{\delta}\mathbf{A} = \mathbf{A}\boldsymbol{\delta} = \mathbf{A}, \tag{1.48}$$

or

$$\delta_{ik}A_{kj} = A_{ij}, \quad A_{ik}\delta_{kj} = A_{ij}. \tag{1.49}$$

The inverse matrix of $\mathbf{A}$ is defined by

$$\mathbf{A}\mathbf{A}^{-1} = \mathbf{A}^{-1}\mathbf{A} = \boldsymbol{\delta}, \tag{1.50}$$

or

$$A_{ij}A_{jk}^{-1} = \delta_{ik}, \quad A_{ij}^{-1}A_{jk} = \delta_{ik}. \tag{1.51}$$

An orthogonal matrix is a matrix that satisfies

$$\mathbf{A}^{-1} = \mathbf{A}^{T}. \tag{1.52}$$

Then

$$\mathbf{A}\mathbf{A}^{T} = \mathbf{A}^{T}\mathbf{A} = \boldsymbol{\delta}, \tag{1.53}$$

or

$$\begin{aligned} A_{ij}A_{jk}^{T} &= A_{ij}A_{kj} = \delta_{ik}, \\ A_{ij}^{T}A_{jk} &= A_{ji}A_{jk} = \delta_{ik}. \end{aligned} \tag{1.54}$$

A vector may be represented by a column matrix as

$$\mathbf{a} = \begin{bmatrix} a_1 \\ a_2 \\ a_3 \end{bmatrix}. \tag{1.55}$$

Then $\mathbf{a}^T$ is a row matrix

$$\mathbf{a}^T = \begin{bmatrix} a_1 & a_2 & a_3 \end{bmatrix}. \tag{1.56}$$

For

$$\mathbf{b} = \mathbf{A}\mathbf{a}, \tag{1.57}$$

or

$$\begin{bmatrix} b_1 \\ b_2 \\ b_3 \end{bmatrix} = \begin{bmatrix} A_{11} & A_{12} & A_{13} \\ A_{21} & A_{22} & A_{23} \\ A_{31} & A_{32} & A_{33} \end{bmatrix} \begin{bmatrix} a_1 \\ a_2 \\ a_3 \end{bmatrix}, \tag{1.58}$$

we can write

$$b_i = \sum_{k=1}^{3} A_{ik} a_k = A_{ik} a_k. \tag{1.59}$$

Obviously, since

$$\boldsymbol{\delta}\mathbf{a} = \mathbf{a}, \tag{1.60}$$

or

$$\begin{bmatrix} 1 & 0 & 0 \\ 0 & 1 & 0 \\ 0 & 0 & 1 \end{bmatrix} \begin{bmatrix} a_1 \\ a_2 \\ a_3 \end{bmatrix} = \begin{bmatrix} a_1 \\ a_2 \\ a_3 \end{bmatrix}, \tag{1.61}$$

we have

$$(\boldsymbol{\delta}\mathbf{a})_i = \delta_{ik} a_k = a_i. \tag{1.62}$$

If

$$\begin{aligned} &\mathbf{A}\mathbf{a} = \lambda\mathbf{a}, \quad A_{ij} a_j = \lambda a_i, \\ &\mathbf{a} \neq 0, \end{aligned} \tag{1.63}$$

then λ is an eigenvalue of $\mathbf{A}$, and $\mathbf{a}$ is the corresponding eigenvector. Then

$$(\mathbf{A} - \lambda\mathbf{1})\mathbf{a} = 0, \quad (A_{ij} - \lambda\delta_{ij})a_j = 0. \tag{1.64}$$

The determinant of the coefficient matrix of the above equation has to vanish, which leads to the so-called characteristic equation

$$\det(\mathbf{A} - \lambda\mathbf{1}) = 0, \quad \det(A_{ij} - \lambda\delta_{ij}) = 0. \tag{1.65}$$

It is a cubic equation for λ. The well-known Cayley-Hamilton theorem states that $\mathbf{A}$ satisfies its characteristic equation. For a real symmetric matrix, the eigenvalues are real, and the eigenvectors corresponding to different eigenvalues are orthogonal to each other.

A square matrix is positive definite if

$$A_{ij}a_i a_j > 0 \quad \text{for any } \mathbf{a} \neq 0. \tag{1.66}$$

A real and invertible matrix can be decomposed into the products of

$$\begin{aligned} \mathbf{A} &= \mathbf{RU} = \mathbf{VR}, \\ A_{ij} &= R_{ik}U_{kj} = V_{ik}R_{kj}, \end{aligned} \tag{1.67}$$

where $\mathbf{U}$ and $\mathbf{V}$ are positive definite symmetric matrices with the same eigenvalues but different eigenvectors, and $\mathbf{R}$ is orthogonal. Equation (1.67) is known as the polar decomposition of $\mathbf{A}$.

1.4 Determinants

We consider 3×3 determinants only. Let

$$\begin{aligned} A = \det(\mathbf{A}) &= \begin{vmatrix} A_{11} & A_{12} & A_{13} \\ A_{21} & A_{22} & A_{23} \\ A_{31} & A_{32} & A_{33} \end{vmatrix} \\ &= A_{11}A_{22}A_{33} + A_{21}A_{32}A_{13} + A_{31}A_{12}A_{23} \\ &\quad - A_{13}A_{22}A_{31} - A_{12}A_{21}A_{33} - A_{11}A_{23}A_{32}. \end{aligned} \tag{1.68}$$

We have

$$\begin{aligned}\varepsilon_{ijk}A_{i1}A_{j2}A_{k3} &= \varepsilon_{123}A_{11}A_{22}A_{33} + \varepsilon_{231}A_{21}A_{32}A_{13} + \varepsilon_{312}A_{31}A_{12}A_{23} \\ &\quad + \varepsilon_{321}A_{31}A_{22}A_{13} + \varepsilon_{213}A_{21}A_{12}A_{33} + \varepsilon_{132}A_{11}A_{32}A_{23} \\ &= A_{11}A_{22}A_{33} + A_{21}A_{32}A_{13} + A_{31}A_{12}A_{23} \\ &\quad - A_{31}A_{22}A_{13} - A_{21}A_{12}A_{33} - A_{11}A_{32}A_{23} = A,\end{aligned} \tag{1.69}$$

or

$$A = \varepsilon_{ijk}A_{i1}A_{j2}A_{k3}. \tag{1.70}$$

Similarly,

$$A = \varepsilon_{ijk}A_{1i}A_{2j}A_{3k}. \tag{1.71}$$

Then the following can be verified:

$$\begin{aligned}\varepsilon_{ijk}A_{il}A_{jm}A_{kn} &= \varepsilon_{lmn}A, \\ \varepsilon_{lmn}A_{il}A_{jm}A_{kn} &= \varepsilon_{ijk}A.\end{aligned} \tag{1.72}$$

Multiplying both sides of Eq. $(1.72)_1$ by A_{lr}^{-1}, we have

$$A_{lr}^{-1}\varepsilon_{ijk}A_{il}A_{jm}A_{kn} = A_{lr}^{-1}\varepsilon_{lmn}A. \tag{1.73}$$

Then

$$\varepsilon_{ijk}A_{il}A_{lr}^{-1}A_{jm}A_{kn} = \varepsilon_{lmn}AA_{lr}^{-1}, \tag{1.74}$$

$$\varepsilon_{ijk}\delta_{ir}A_{jm}A_{kn} = \varepsilon_{lmn}AA_{lr}^{-1}, \tag{1.75}$$

$$\varepsilon_{rjk}A_{jm}A_{kn} = \varepsilon_{lmn}AA_{lr}^{-1}. \tag{1.76}$$

Replacing the index r by i in Eq. (1.76), we obtain

$$\varepsilon_{ijk}A_{jm}A_{kn} = A\varepsilon_{lmn}A_{li}^{-1}, \tag{1.77}$$

which will be useful later.

If we multiply both sides of Eq. (1.77) by ε_{pmn}, we have

$$\begin{aligned}\varepsilon_{pmn}\varepsilon_{ijk}A_{jm}A_{kn} &= A\varepsilon_{pmn}\varepsilon_{lmn}A_{li}^{-1} = A\varepsilon_{mnl}\varepsilon_{mnp}A_{li}^{-1} \\ &= A\left(\delta_{nn}\delta_{lp} - \delta_{np}\delta_{ln}\right)A_{li}^{-1} = A\left(3\delta_{lp} - \delta_{lp}\right)A_{li}^{-1} \\ &= A2\delta_{lp}A_{li}^{-1} = 2AA_{pi}^{-1},\end{aligned} \tag{1.78}$$

where the ε–δ identity in Eq. (1.27) has been used. Replacing the index p by l in Eq. (1.78), we obtain

$$\varepsilon_{ijk}\varepsilon_{lmn}A_{jm}A_{kn} = 2AA_{li}^{-1}, \tag{1.79}$$

which will also be useful later. It can also be shown using Eq. (1.72) that

$$A = \frac{1}{6}\varepsilon_{ijk}\varepsilon_{lmn}A_{il}A_{jm}A_{kn}. \tag{1.80}$$

1.5 Scalar and Vector Fields

In this section, we consider scalar and vector fields as functions of $\mathbf{x}$. Let

$$\varphi = \varphi(\mathbf{x}), \quad \mathbf{a} = \mathbf{a}(\mathbf{x}). \tag{1.81}$$

We introduce the comma convention for partial derivatives with respect to the spatial coordinates by

$$\frac{\partial(\ )}{\partial x_i} = \partial_i(\ ) = (\ )_{,i}. \tag{1.82}$$

Then the gradient operator can be written as

$$\nabla = \mathbf{e}_1\frac{\partial}{\partial x_1} + \mathbf{e}_2\frac{\partial}{\partial x_2} + \mathbf{e}_3\frac{\partial}{\partial x_3} = \mathbf{e}_i\partial_i. \tag{1.83}$$

The gradient of a scalar field can be written as

$$\nabla\varphi = \frac{\partial\varphi}{\partial x_1}\mathbf{e}_1 + \frac{\partial\varphi}{\partial x_2}\mathbf{e}_2 + \frac{\partial\varphi}{\partial x_3}\mathbf{e}_3 = \sum_{i=1}^{3}\frac{\partial\varphi}{\partial x_i}\mathbf{e}_i = \mathbf{e}_i\partial_i\varphi = \varphi_{,i}\mathbf{e}_i, \tag{1.84}$$

or

$$(\nabla\varphi)_i = \varphi_{,i}. \tag{1.85}$$

The divergence of a vector field is given by

$$\begin{aligned}\nabla\cdot\mathbf{a} &= \frac{\partial a_1}{\partial x_1}+\frac{\partial a_2}{\partial x_2}+\frac{\partial a_3}{\partial x_3}=\sum_{i=1}^{3}\frac{\partial a_i}{\partial x_i}\\ &= a_{1,1}+a_{2,2}+a_{3,3}\\ &= \mathbf{e}_i\partial_i\cdot\mathbf{a}=\mathbf{e}_i\partial_i\cdot a_j\mathbf{e}_j=a_{j,i}\mathbf{e}_i\cdot\mathbf{e}_j=a_{j,i}\delta_{ij}=a_{i,i},\end{aligned} \tag{1.86}$$

or

$$\nabla\cdot\mathbf{a}=a_{i,i}. \tag{1.87}$$

The curl (rotation) of a vector field takes the following form:

$$\begin{aligned}\nabla\times\mathbf{a} &= \mathbf{e}_i\partial_i\times\mathbf{a}=\mathbf{e}_i\partial_i\times a_j\mathbf{e}_j=a_{j,i}\mathbf{e}_i\times\mathbf{e}_j\\ &= \varepsilon_{kij}a_{j,i}\mathbf{e}_k=\varepsilon_{kij}\partial_i a_j\mathbf{e}_k,\end{aligned} \tag{1.88}$$

or

$$\nabla\times\mathbf{a}=\varepsilon_{kij}a_{j,i}\mathbf{e}_k,\quad (\nabla\times\mathbf{a})_k=\varepsilon_{kij}a_{j,i}. \tag{1.89}$$

The Laplace operator (Laplacian) is defined by

$$\begin{aligned}\nabla^2 &= \frac{\partial^2}{\partial x_1^2}+\frac{\partial^2}{\partial x_2^2}+\frac{\partial^2}{\partial x_3^2}=\nabla\cdot\nabla\\ &= \mathbf{e}_i\partial_i\cdot\mathbf{e}_j\partial_j=\mathbf{e}_i\cdot\mathbf{e}_j\partial_i\partial_j=\delta_{ij}\partial_i\partial_j=\partial_i\partial_i.\end{aligned} \tag{1.90}$$

Then

$$\nabla^2\varphi=\varphi_{,ii}=\frac{\partial^2\varphi}{\partial x_1^2}+\frac{\partial^2\varphi}{\partial x_2^2}+\frac{\partial^2\varphi}{\partial x_3^2}, \tag{1.91}$$

$$\nabla^2\mathbf{a}=\partial_i\partial_i a_j\mathbf{e}_j=a_{j,ii}\mathbf{e}_j,\quad \left(\nabla^2\mathbf{a}\right)_j=a_{j,ii}. \tag{1.92}$$

As an example, we prove the following vector identity [3]:

$$\nabla\times(\nabla\times\mathbf{v})=\nabla(\nabla\cdot\mathbf{v})-\nabla^2\mathbf{v}. \tag{1.93}$$

Let

$$\mathbf{u} = \nabla \times \mathbf{v}, \quad u_k = \varepsilon_{klm}\partial_l v_m = \varepsilon_{klm}v_{m,l}. \tag{1.94}$$

We have

$$\begin{aligned}
[\nabla \times (\nabla \times \mathbf{v})]_i &= [\nabla \times \mathbf{u}]_i \\
&= \varepsilon_{ijk}\partial_j u_k = \varepsilon_{ijk}u_{k,j} = \varepsilon_{ijk}(\varepsilon_{klm}v_{m,l})_{,j} \\
&= \varepsilon_{ijk}\varepsilon_{klm}v_{m,lj} = \varepsilon_{kij}\varepsilon_{klm}v_{m,lj} \\
&= (\delta_{il}\delta_{jm} - \delta_{im}\delta_{jl})v_{m,lj} = v_{j,ij} - v_{i,jj} \\
&= v_{j,ji} - v_{i,jj} = [\nabla(\nabla \cdot \mathbf{v})]_i - (\nabla^2\mathbf{v})_i,
\end{aligned} \tag{1.95}$$

where the ε–δ identity in Eq. (1.27) has been used.

For a vector field over a body v with a boundary surface s whose outward unit normal is $\mathbf{n}$, the divergence theorem [3] can be written as

$$\int_s \mathbf{n} \cdot \mathbf{a} ds = \int_v \nabla \cdot \mathbf{a} dv, \tag{1.96}$$

or

$$\int_s n_i a_i ds = \int_v a_{i,i} dv. \tag{1.97}$$

1.6 Tensors

The concept of tensors is closely related to coordinate transformations. Consider the two Cartesian coordinate systems in Fig. 1.2. They are at the same point with different orientations.

The transformation coefficients between the two coordinate systems are defined by

$$\mathbf{e}'_i \cdot \mathbf{e}_j = \alpha_{ij}. \tag{1.98}$$

If we write

$$\mathbf{e}'_i = \lambda_{ij}\mathbf{e}_j, \tag{1.99}$$

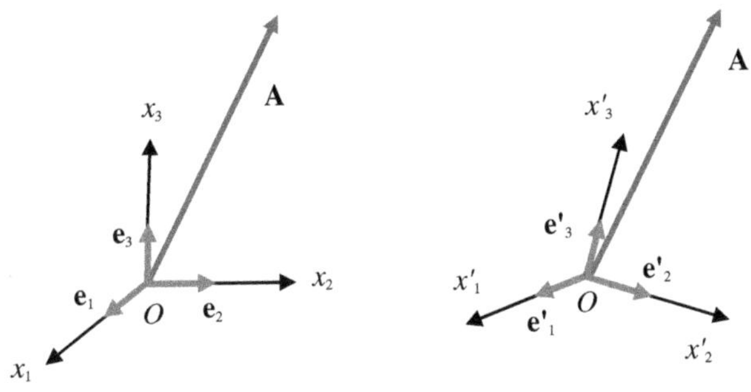

Fig. 1.2. Two Cartesian coordinate systems and a vector.

then, by dotting both sides with $\mathbf{e}_k$, we have

$$\mathbf{e}'_i \cdot \mathbf{e}_k = \lambda_{ij}\mathbf{e}_j \cdot \mathbf{e}_k, \tag{1.100}$$

$$\alpha_{ik} = \lambda_{ij}\delta_{jk} = \lambda_{ik}. \tag{1.101}$$

Therefore,

$$\mathbf{e}'_i = \alpha_{ij}\mathbf{e}_j, \tag{1.102}$$

which is obvious from the definition of α_{ij}.

Both coordinate systems are assumed to be orthogonal and their basis vectors are orthonormal. Hence,

$$\delta_{ij} = \mathbf{e}'_i \cdot \mathbf{e}'_j = \alpha_{ik}\mathbf{e}_k \cdot \alpha_{jl}\mathbf{e}_l = \alpha_{ik}\alpha_{jl}\mathbf{e}_k \cdot \mathbf{e}_l = \alpha_{ik}\alpha_{jl}\delta_{kl} = \alpha_{ik}\alpha_{jk}, \tag{1.103}$$

which shows that α_{ij} form an orthogonal matrix satisfying

$$\alpha_{ik}\alpha_{jk} = \delta_{ij}. \tag{1.104}$$

Similarly, it can be shown that

$$\alpha_{ki}\alpha_{kj} = \delta_{ij}. \tag{1.105}$$

A scalar has the same value in any coordinate system. A vector $\mathbf{b}$ has different components in different coordinate systems but the vector itself as a whole is coordinate independent under a coordinate transformation, i.e.,

$$\mathbf{b}' = b'_i\mathbf{e}'_i = b_i\mathbf{e}_i = \mathbf{b}. \tag{1.106}$$

To see the implications of Eq. (1.106) on the components of the vector, we dot both sides of the equation by $\mathbf{e}'_j$ and obtain

$$b'_i \mathbf{e}'_i \cdot \mathbf{e}'_j = b_i \mathbf{e}_i \cdot \mathbf{e}'_j, \tag{1.107}$$

$$b'_i \delta_{ij} = b_i \mathbf{e}'_j \cdot \mathbf{e}_i, \tag{1.108}$$

$$b'_j = \alpha_{ji} b_i. \tag{1.109}$$

Multiplying both sides by α_{jk}, we have

$$\alpha_{jk} b'_j = \alpha_{jk} \alpha_{ji} b_i, \tag{1.110}$$

$$\alpha_{jk} b'_j = \delta_{ki} b_i, \tag{1.111}$$

$$\alpha_{jk} b'_j = b_k. \tag{1.112}$$

In summary, for the coordinate independence of $\mathbf{b}$, its components transform according to

$$b'_i = \alpha_{ij} b_j, \quad b_i = \alpha_{ji} b'_j. \tag{1.113}$$

In the matrix notation of Section 1.3, for the components of $\mathbf{b}$ as a column matrix, Eq. $(1.113)_1$ may be written as $[b'_i] = [\alpha_{im}][b_m]$ or $\mathbf{b}' = \boldsymbol{\alpha}\mathbf{b}$. In particular, for the position vector $\mathbf{x}$, we have

$$x'_i = \alpha_{ij} x_j, \quad x_i = \alpha_{ji} x'_j. \tag{1.114}$$

In general, scalars and vectors are considered as zero- and first-order tensors, respectively. The definitions of the second- and higher-order tensors are based on the transformation behaviors of their components:

	Transformation under $x'_i = \alpha_{ij} x_j$	
Zero order (scalar)	$\varphi' = \varphi$	
First order (vector)	$b'_i = \alpha_{ij} b_j$	
Second order	$A'_{ij} = \alpha_{im}\alpha_{jn} A_{mn}$	(1.115)
Third order	$A'_{ijk} = \alpha_{im}\alpha_{jn}\alpha_{kp} A_{mnp}$	
Fourth order	$A'_{ijkl} = \alpha_{im}\alpha_{jn}\alpha_{kp}\alpha_{lq} A_{mnpq}$	
......		

The dyadic (tensor) product of two vectors is denoted by

$$a_i\mathbf{e}_i \otimes b_j\mathbf{e}_j = a_ib_j\mathbf{e}_i \otimes \mathbf{e}_j = a_ib_j\mathbf{e}_i\mathbf{e}_j, \tag{1.116}$$

where $\otimes$ is usually omitted. A second-order tensor can be written as

$$\mathbf{A} = A_{ij}\mathbf{e}_i \otimes \mathbf{e}_j = A_{ij}\mathbf{e}_i\mathbf{e}_j. \tag{1.117}$$

The coordinate independence of a second-order tensor $\mathbf{A}$ can be seen as follows:

$$\begin{aligned}\mathbf{A}' &= A'_{ij}\mathbf{e}'_i\mathbf{e}'_j = (\alpha_{ik}\alpha_{jl}A_{kl})(\alpha_{im}\mathbf{e}_m)(\alpha_{jn}\mathbf{e}_n)\\ &= \alpha_{ik}\alpha_{im}\alpha_{jl}\alpha_{jn}A_{kl}\mathbf{e}_m\mathbf{e}_n = \delta_{km}\delta_{ln}A_{kl}\mathbf{e}_m\mathbf{e}_n\\ &= A_{mn}\mathbf{e}_m\mathbf{e}_n = \mathbf{A}.\end{aligned} \tag{1.118}$$

Although the above $\mathbf{A}$ as a tensor is coordinate independent, its components change according to $A'_{ij} = \alpha_{im}\alpha_{jn}A_{mn}$ which, in the matrix notation of Section 1.3, may be written as $[A'_{ij}] = [\alpha_{im}][A_{mn}][\alpha^T_{nj}]$ or simply $\mathbf{A}' = \boldsymbol{\alpha}\mathbf{A}\boldsymbol{\alpha}^T$. A simple example of a second-order tensor is

$$\delta'_{ij} = \mathbf{e}'_i \cdot \mathbf{e}'_j = (\alpha_{im}\mathbf{e}_m)\cdot(\alpha_{jn}\mathbf{e}_n) = \alpha_{im}\alpha_{jn}\mathbf{e}_m \cdot \mathbf{e}_n = \alpha_{im}\alpha_{jn}\delta_{mn}. \tag{1.119}$$

An example of a third-order tensor is

$$\begin{aligned}\varepsilon'_{ijk} &= \mathbf{e}'_i \cdot (\mathbf{e}'_j \times \mathbf{e}'_k) = \alpha_{il}\mathbf{e}_l \cdot (\alpha_{jm}\mathbf{e}_m \times \alpha_{kn}\mathbf{e}_n)\\ &= \alpha_{il}\alpha_{jm}\alpha_{kn}\mathbf{e}_l \cdot (\mathbf{e}_m \times \mathbf{e}_n) = \alpha_{il}\alpha_{jm}\alpha_{kn}\varepsilon_{lmn}.\end{aligned} \tag{1.120}$$

For tensor operations, we use second-order tensors as examples below. Higher-order tensors are similar. A tensor is a zero tensor if all of its components are zero. For

$$\mathbf{A} = A_{ij}\mathbf{e}_i\mathbf{e}_j, \quad \mathbf{B} = B_{ij}\mathbf{e}_i\mathbf{e}_j, \tag{1.121}$$

the equality means the following:

$$\mathbf{A} = \mathbf{B} \Leftrightarrow A_{ij} = B_{ij}. \tag{1.122}$$

The addition of two tensors and the multiplication of a tensor by a scalar are defined by

$$(\mathbf{A}+\mathbf{B})_{ij} = A_{ij} + B_{ij}, \tag{1.123}$$

$$(\lambda\mathbf{A})_{ij} = \lambda A_{ij}. \tag{1.124}$$

The dot product of two tensors is given by

$$\begin{aligned}\mathbf{A}\cdot\mathbf{B} &= (A_{ij}\mathbf{e}_i\mathbf{e}_j)\cdot(B_{kl}\mathbf{e}_k\mathbf{e}_l) = A_{ij}B_{kl}\mathbf{e}_i\mathbf{e}_j\cdot\mathbf{e}_k\mathbf{e}_l\\ &= A_{ij}B_{kl}\mathbf{e}_i\delta_{jk}\mathbf{e}_l = A_{ik}B_{kl}\mathbf{e}_i\mathbf{e}_l,\end{aligned} \tag{1.125}$$

$$(\mathbf{A}\cdot\mathbf{B})_{il} = A_{ik}B_{kl}. \tag{1.126}$$

The tensor product of $\mathbf{A}$ and $\mathbf{B}$ is

$$\begin{aligned}\mathbf{A}\otimes\mathbf{B} &= (A_{ij}\mathbf{e}_i\mathbf{e}_j)\otimes(B_{kl}\mathbf{e}_k\mathbf{e}_l)\\ &= A_{ij}B_{kl}\mathbf{e}_i\mathbf{e}_j\otimes\mathbf{e}_k\mathbf{e}_l = A_{ij}B_{kl}\mathbf{e}_i\mathbf{e}_j\mathbf{e}_k\mathbf{e}_l,\end{aligned} \tag{1.127}$$

$$(\mathbf{A}\otimes\mathbf{B})_{ijkl} = A_{ij}B_{kl}. \tag{1.128}$$

The contraction of a tensor is

$$\mathrm{tr}(\mathbf{A}) = A_{ii}. \tag{1.129}$$

It is often useful to split a tensor $\mathbf{A}$ into the sum of two tensors. One is spherical, which is the product of a scalar and $\boldsymbol{\delta}$. The other is deviatoric with a zero divergence:

$$A_{ij} = S_{ij} + D_{ij}, \tag{1.130}$$

$$\begin{aligned}S_{ij} &= \frac{A_{kk}}{3}\delta_{ij}, \quad S_{ll} = A_{kk},\\ D_{ij} &= A_{ij} - \frac{A_{kk}}{3}\delta_{ij}, \quad D_{ll} = 0.\end{aligned} \tag{1.131}$$

The gradient of a tensor field is given by

$$\nabla\mathbf{A} = (\partial_i\mathbf{e}_i)(A_{jk}\mathbf{e}_j\mathbf{e}_k) = A_{jk,i}\mathbf{e}_i\mathbf{e}_j\mathbf{e}_k, \tag{1.132}$$

or

$$(\nabla\mathbf{A})_{ijk} = \partial_i A_{jk} = A_{jk,i}. \tag{1.133}$$

The divergence of $\mathbf{A}$ is

$$\nabla\cdot\mathbf{A} = (\partial_i\mathbf{e}_i)\cdot(A_{jk}\mathbf{e}_j\mathbf{e}_k) = A_{jk,i}\mathbf{e}_i\cdot\mathbf{e}_j\mathbf{e}_k = A_{jk,i}\delta_{ij}\mathbf{e}_k = A_{jk,j}\mathbf{e}_k, \tag{1.134}$$

or

$$(\nabla\cdot\mathbf{A})_k = A_{jk,j}. \tag{1.135}$$

For comparison, we summarize the tensor notation, the tensor component notation, and the matrix notation of common operations of tensors below [4]:

Tensor	Tensor comp.	Matrix
$\mathbf{a} \cdot \mathbf{b}$	$a_i b_i$	$\mathbf{a}^T \mathbf{b}$
$\mathbf{a} \otimes \mathbf{b}$ or $\mathbf{ab}$	$a_i b_j$	$\mathbf{ab}^T$
$\mathbf{A} \cdot \mathbf{a}$	$A_{ij} a_j$	$\mathbf{Aa}$
$\mathbf{a} \cdot \mathbf{A}$	$a_i A_{ij}$	$\mathbf{a}^T \mathbf{A}$
$\mathbf{a} \cdot \mathbf{A} \cdot \mathbf{b}$	$a_i A_{ij} b_j$	$\mathbf{a}^T \mathbf{Ab}$
$\mathbf{A} \cdot \mathbf{B}$	$A_{ik} B_{kj}$	$\mathbf{AB}$
$\mathbf{A} \cdot \mathbf{B}^T$	$A_{ik} B_{jk}$	$\mathbf{AB}^T$
$\mathbf{A} \cdot \mathbf{B} \cdot \mathbf{C}$	$A_{ik} B_{kl} C_{lj}$	$\mathbf{ABC}$

1.7 Divergence Theorem

The divergence theorem for vector fields can be generalized to tensor fields as

$$\int_S \mathbf{n} \cdot \mathbf{A} ds = \int_v \nabla \cdot \mathbf{A} dv, \tag{1.136}$$

or

$$\int_s n_i A_{ij} ds = \int_v A_{ij,i} dv. \tag{1.137}$$

To prove Eq. (1.137), we begin with the divergence theorem for a vector field $\mathbf{a}$ in Eq. (1.96) [1]. Let

$$\mathbf{a} = f\mathbf{e}_i. \tag{1.138}$$

Then, using Eq. (1.96), we have

$$\int_s n_i f ds = \int_v \frac{\partial f}{\partial x_i} dv. \tag{1.139}$$

For a second-order tensor, let

$$f = A_{kl}. \tag{1.140}$$

Then

$$\int_s n_i A_{kl} ds = \int_v \frac{\partial A_{kl}}{\partial x_i} dv. \tag{1.141}$$

Setting $i = k = 1$ in Eq. (1.141), we have

$$\int_s n_1 A_{1l} ds = \int_v \frac{\partial A_{1l}}{\partial x_1} dv. \tag{1.142}$$

Similarly,

$$\int_s n_2 A_{2l} ds = \int_v \frac{\partial A_{2l}}{\partial x_2} dv, \tag{1.143}$$

$$\int_s n_3 A_{3l} ds = \int_v \frac{\partial A_{3l}}{\partial x_3} dv. \tag{1.144}$$

Adding Eqs. (1.142)–(1.144), we arrive at the divergence theorem for a tensor field:

$$\int_s n_i A_{il} ds = \int_v \frac{\partial A_{il}}{\partial x_i} dv. \tag{1.145}$$

The above divergence theorem assumes that $\mathbf{A}$ is smooth enough in v. Next, consider the case when $\mathbf{A}$ has a discontinuity surface Σ in v as shown in Fig. 1.3 [5]. Across Σ, $\mathbf{A}$ is not differentiable. Σ divides v into two parts. We denote

$$\begin{aligned} &v = v^+ \cup v^-, \quad \partial v = s^+ \cup s^- = s, \\ &\partial v^+ = s^+ \cup \Sigma, \quad \partial v^- = s^- \cup \Sigma. \end{aligned} \tag{1.146}$$

Since $\mathbf{A}$ is sufficiently smooth in each part of v, we apply Eq. (1.136) to each part of v and obtain

$$\int_{s^-} \mathbf{n} \cdot \mathbf{A} ds + \int_\Sigma \mathbf{n} \cdot \mathbf{A}^- ds = \int_{v^-} \nabla \cdot \mathbf{A} dv, \tag{1.147}$$

$$\int_{s^+} \mathbf{n} \cdot \mathbf{A} ds + \int_\Sigma (-\mathbf{n}) \cdot \mathbf{A}^+ ds = \int_{v^+} \nabla \cdot \mathbf{A} dv. \tag{1.148}$$

Adding Eqs. (1.147) and (1.148), we obtain the generalized divergence theorem below when a discontinuity surface is present:

$$\int_s \mathbf{n} \cdot \mathbf{A} ds = \int_{v-\Sigma} \nabla \cdot \mathbf{A} dv + \int_\Sigma \mathbf{n} \cdot [\mathbf{A}] ds, \tag{1.149}$$

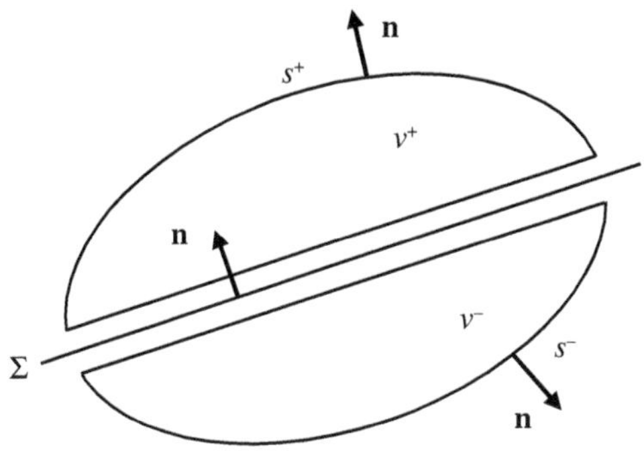

Fig. 1.3. A region v in which $\mathbf{A}$ has a discontinuity surface Σ.

where

$$[\mathbf{A}] = \mathbf{A}^{+} - \mathbf{A}^{-}, \tag{1.150}$$

and the integration over v–Σ is simply the sum of the integrations over v_1 and v_2. It does not have the contribution from Σ. When $\mathbf{n} \cdot \mathbf{A}$ is continuous across Σ, Eq. (1.149) reduces to the conventional divergence theorem in Eq. (1.136).

1.8 Isotropic Tensors

In dealing with isotropic materials, we need the concept of isotropic tensors whose components are all scalars. For example, a second-order isotropic tensor satisfies

$$A'_{ij} = A_{ij}. \tag{1.151}$$

It can be shown [1] that all zero-order tensors (scalars) are isotropic. The only first-order isotropic tensor (vector) is the zero vector. All second-order isotropic tensors have the form of $\lambda\delta_{ij}$. All third-order isotropic tensors have the form of $\lambda\varepsilon_{ijk}$. All four-order isotropic tensors have the form of

$$A_{ijkl} = \lambda\delta_{ij}\delta_{kl} + \alpha\delta_{ik}\delta_{jl} + \beta\delta_{il}\delta_{jk}. \tag{1.152}$$

We prove the case of second-order tensors below. Obviously, $\lambda\delta_{ij}$ is isotropic. To prove that all second-order isotropic tensors are of the form of $\lambda\delta_{ij}$, consider the rotation of the three axes along the three directions of $(1, 2, 3)$ into $(3, 1, 2)$ with

$$[\alpha_{ij}] = \begin{bmatrix} 0 & 1 & 0 \\ 0 & 0 & 1 \\ 1 & 0 & 0 \end{bmatrix}. \tag{1.153}$$

Then, from the isotropy of $\mathbf{A}$ and its transformation rule, we have

$$\begin{aligned} A_{11} &= A'_{11} = \alpha_{1k}\alpha_{1l}A_{kl} = \alpha_{12}\alpha_{12}A_{22} = A_{22}, \\ A_{22} &= A'_{22} = \alpha_{2k}\alpha_{2l}A_{kl} = \alpha_{23}\alpha_{23}A_{33} = A_{33}. \end{aligned} \tag{1.154}$$

Hence,

$$A_{11} = A_{22} = A_{33} = \lambda. \tag{1.155}$$

At the same time,

$$\begin{aligned} A_{12} &= A'_{12} = \alpha_{1k}\alpha_{2l}A_{kl} = \alpha_{12}\alpha_{23}A_{23} = A_{23}, \\ A_{23} &= A'_{23} = \alpha_{2k}\alpha_{3l}A_{kl} = \alpha_{23}\alpha_{31}A_{31} = A_{31}. \end{aligned} \tag{1.156}$$

Hence,

$$A_{12} = A_{23} = A_{31}. \tag{1.157}$$

Similarly,

$$\begin{aligned} A_{21} &= A'_{21} = \alpha_{2k}\alpha_{1l}A_{kl} = \alpha_{23}\alpha_{12}A_{32} = A_{32}, \\ A_{32} &= A'_{32} = \alpha_{3k}\alpha_{2l}A_{kl} = \alpha_{31}\alpha_{23}A_{13} = A_{13}. \end{aligned} \tag{1.158}$$

Hence,

$$A_{21} = A_{32} = A_{13}. \tag{1.159}$$

Then consider the rotation about the x_3 axis by 180° for which

$$[\alpha_{ij}] = \begin{bmatrix} -1 & 0 & 0 \\ 0 & -1 & 0 \\ 0 & 0 & 1 \end{bmatrix}. \tag{1.160}$$

Under Eq. (1.160), we have

$$\begin{aligned} A_{23} &= A'_{23} = \alpha_{2k}\alpha_{3l}A_{kl} = \alpha_{22}\alpha_{33}A_{23} = -A_{23}, \\ A_{32} &= A'_{32} = \alpha_{3k}\alpha_{2l}A_{kl} = \alpha_{33}\alpha_{22}A_{32} = -A_{32}. \end{aligned} \tag{1.161}$$

Hence,

$$A_{23} = 0, \quad A_{32} = 0. \tag{1.162}$$

For a fourth-order isotropic tensor, if it is also symmetric with

$$A_{ijkl} = A_{jikl}, \tag{1.163}$$

then, from Eq. (1.152),

$$\lambda\delta_{ij}\delta_{kl} + \alpha\delta_{ik}\delta_{jl} + \beta\delta_{il}\delta_{jk} = \lambda\delta_{ji}\delta_{kl} + \alpha\delta_{jk}\delta_{il} + \beta\delta_{jl}\delta_{ik}, \tag{1.164}$$

or

$$\alpha\delta_{ik}\delta_{jl} + \beta\delta_{il}\delta_{jk} = \alpha\delta_{jk}\delta_{il} + \beta\delta_{jl}\delta_{ik}. \tag{1.165}$$

Setting $i = k = 1$ and $j = l = 2$ in Eq. (1.165), we obtain

$$\alpha\delta_{11}\delta_{22} + \beta\delta_{12}\delta_{21} = \alpha\delta_{21}\delta_{12} + \beta\delta_{22}\delta_{11}, \tag{1.166}$$

which leads to

$$\alpha = \beta. \tag{1.167}$$

Then

$$A_{ijkl} = \lambda\delta_{ij}\delta_{kl} + \mu\,(\delta_{ik}\delta_{jl} + \delta_{il}\delta_{jk}), \tag{1.168}$$

where we have denoted α or β by μ.

1.9 Rotationally Invariant Tensor Functions

A rotation of a Cartesian coordinate system is described by an orthogonal tensor $\mathbf{Q}$ through

$$\begin{aligned} &x'_i = Q_{ij}x_j \quad \mathbf{x}' = \mathbf{Qx}, \\ &Q_{ik}Q_{jk} = Q_{ki}Q_{kj} = \delta_{ij}, \\ &\mathbf{QQ}^T = \mathbf{Q}^T\mathbf{Q} = \mathbf{1}, \quad \det(\mathbf{Q}) = 1, \end{aligned} \tag{1.169}$$

where the operations involving the boldface $\mathbf{Q}$ are based on the matrix notation in Section 1.3 for the components of the relevant vectors and tensors, rather than the tensor notation and operation. The same is true for the rest of this section. A scalar ψ as a function of vectors and second-order tensors, in the special case of one vector $\mathbf{v}$ and one tensor A, may be written as

$$\psi = f(v_i, A_{ij}). \tag{1.170}$$

As a scalar, ψ does not change its value under $\mathbf{Q}$, i.e.,

$$\psi' = \psi. \tag{1.171}$$

f is said to be form-invariant under $\mathbf{Q}$ if it satisfies

$$f(v_i', A_{ij}') = f(v_i, A_{ij}), \tag{1.172}$$

$$\begin{aligned} f(Q_{ij}v_j, Q_{ik}Q_{jl}A_{kl}) &= f(v_i, A_{ij}), \\ f(\mathbf{Qv}, \mathbf{QAQ}^T) &= f(\mathbf{v}, \mathbf{A}). \end{aligned} \tag{1.173}$$

Consider the simplest case of a scalar function f of a two-dimensional vector $\mathbf{v}$ first. A two-dimensional rotation is described by

$$\mathbf{Q} = \begin{bmatrix} \cos\theta & -\sin\theta \\ \sin\theta & \cos\theta \end{bmatrix}, \quad \mathbf{QQ}^T = \mathbf{1}, \quad \det(\mathbf{Q}) = 1. \tag{1.174}$$

Under $\mathbf{Q}$,

$$\begin{aligned} \mathbf{v}' = \begin{bmatrix} v_1' \\ v_2' \end{bmatrix} = \mathbf{Qv} &= \begin{bmatrix} \cos\theta & -\sin\theta \\ \sin\theta & \cos\theta \end{bmatrix} \begin{bmatrix} v_1 \\ v_2 \end{bmatrix} \\ &= \begin{bmatrix} v_1\cos\theta - v_2\sin\theta \\ v_1\sin\theta + v_2\cos\theta \end{bmatrix}, \end{aligned} \tag{1.175}$$

as shown in Fig. 1.4.

It is easy to verify that the magnitude of a vector $\mathbf{v}$ is a form-invariant scalar function of the components of the vector. In general, the functional form or structure of f, a rotationally form-invariant scalar function of $\mathbf{v}$, must satisfy

$$f(\mathbf{v}) = f(\mathbf{v}'), \tag{1.176}$$

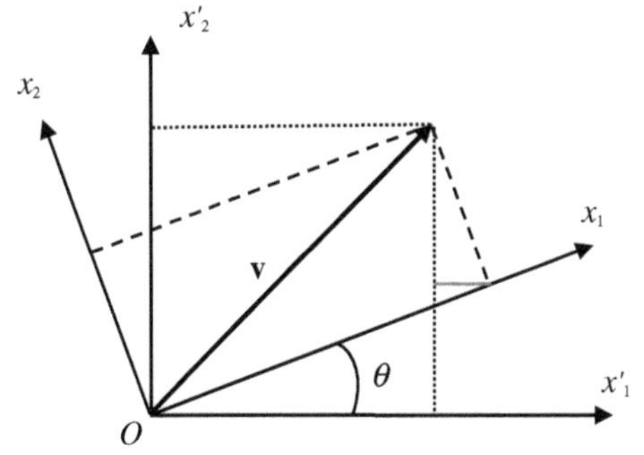

Fig. 1.4. A vector **v** in two rectangular coordinate systems.

or

$$f(v_1, v_2) = f(v_1', v_2'), \tag{1.177}$$

$$f(v_1, v_2) = f(v_1 \cos\theta - v_2 \sin\theta, v_1 \sin\theta + v_2 \cos\theta). \tag{1.178}$$

Differentiating both sides of Eq. (1.178) with respect to θ, we obtain

$$0 = \frac{\partial f}{\partial v_1'}(-v_1 \sin\theta - v_2 \cos\theta) + \frac{\partial f}{\partial v_2'}(v_1 \cos\theta - v_2 \sin\theta), \tag{1.179}$$

or

$$-v_2' \frac{\partial f}{\partial v_1'} + v_1' \frac{\partial f}{\partial v_2'} = 0, \tag{1.180}$$

which is a first-order linear and homogeneous partial differential equation for f. Its characteristic equation is

$$\frac{dv_1'}{-v_2'} = \frac{dv_2'}{v_1'}, \tag{1.181}$$

or

$$v_1' dv_1' + v_2' dv_2' = 0. \tag{1.182}$$

A first integral of Eq. (1.182) is

$$(v_1')^2 + (v_2')^2 = \mathbf{v}' \cdot \mathbf{v}' = C. \tag{1.183}$$

Then the general solution of Eq. (1.180) can be written as

$$f = f[(v_1')^2 + (v_2')^2] = f(\mathbf{v}' \cdot \mathbf{v}') = f(\mathbf{v} \cdot \mathbf{v}). \tag{1.184}$$

In the case when f is a function of two vectors, $\mathbf{u}$ and $\mathbf{v}$, it is easy to show that their magnitudes and inner product as well as the angle between them are form-invariant scalar functions of the components of the vectors. In general, for rotational invariance, f must satisfy

$$f(\mathbf{u}, \mathbf{v}) = f(\mathbf{u}', \mathbf{v}'), \tag{1.185}$$

or

$$f(u_1, u_2, v_1, v_2) = f(u_1', u_2', v_1', v_2'), \tag{1.186}$$

$$\begin{aligned} f(u_1, u_2, v_1, v_2) = f(&u_1 \cos\theta - u_2 \sin\theta, u_1 \sin\theta + u_2 \cos\theta, \\ &v_1 \cos\theta - v_2 \sin\theta, v_1 \sin\theta + v_2 \cos\theta). \end{aligned} \tag{1.187}$$

Differentiating both sides of Eq. (1.187) with respect to θ, we obtain

$$\begin{aligned} 0 = {}& \frac{\partial f}{\partial u_1'}(-u_1 \sin\theta - u_2 \cos\theta) + \frac{\partial f}{\partial u_2'}(u_1 \cos\theta - u_2 \sin\theta) \\ &+ \frac{\partial f}{\partial v_1'}(-v_1 \sin\theta - v_2 \cos\theta) + \frac{\partial f}{\partial v_2'}(v_1 \cos\theta - v_2 \sin\theta), \end{aligned} \tag{1.188}$$

or

$$-u_2' \frac{\partial f}{\partial u_1'} + u_1' \frac{\partial f}{\partial u_2'} - v_2' \frac{\partial f}{\partial v_1'} + v_1' \frac{\partial f}{\partial v_2'} = 0. \tag{1.189}$$

The characteristic equations of Eq. (1.189) are

$$\frac{du_1'}{-u_2'} = \frac{du_2'}{u_1'} = \frac{dv_1'}{-v_2'} = \frac{dv_2'}{v_1'}. \tag{1.190}$$

Three of the first integrals are

$$\begin{aligned} (u_1')^2 + (u_2')^2 &= \mathbf{u}' \cdot \mathbf{u}' = C_1, \\ (v_1')^2 + (v_2')^2 &= \mathbf{v}' \cdot \mathbf{v}' = C_2, \\ u_1'v_1' + u_2'v_2' &= \mathbf{u}' \cdot \mathbf{v}' = C_3. \end{aligned} \tag{1.191}$$

Then the general solution of Eq. (1.189) is

$$f = f(\mathbf{u}' \cdot \mathbf{u}', \mathbf{v}' \cdot \mathbf{v}', \mathbf{u}' \cdot \mathbf{v}') = f(\mathbf{u} \cdot \mathbf{u}, \mathbf{v} \cdot \mathbf{v}, \mathbf{u} \cdot \mathbf{v}). \tag{1.192}$$

Thus, f is a function of the three inner products of $\mathbf{u}$ and $\mathbf{v}$. There are other first integrals, e.g.,

$$u_1'v_2' - u_2'v_1' = \begin{vmatrix} u_1' & u_2' \\ v_1' & v_2' \end{vmatrix} = C_4, \tag{1.193}$$

which is not independent to the ones in Eq. (1.191):

$$(u_1'v_2' - u_2'v_1')^2 = [(u_1')^2 + (u_2')^2][(v_1')^2 + (v_2')^2] - (u_1'v_1' + u_2'v_2')^2. \tag{1.194}$$

The sign of $u_1'v_2' - u_2'v_1'$ depends on whether $\mathbf{u}$, $\mathbf{v}$ and $\mathbf{e}_3$ is right-handed or not, where $\mathbf{e}_3$ is perpendicular to the plane of $\mathbf{u}$ and $\mathbf{v}$. In the case of a group of three-dimensional vectors, according to a theorem of Cauchy, the rotational invariance is assured when f is a function of the inner products of the vectors and the determinants formed by any three vectors of the group [5].

A rotationally form-invariant scalar function f of a tensor $\mathbf{A}$ satisfies

$$\begin{aligned} f(A_{ij}) &= f(A'_{ij}) = f(Q_{ik}Q_{jl}A_{kl}), \\ f(\mathbf{A}) &= f(\mathbf{A}') = f(\mathbf{QAQ}^T). \end{aligned} \tag{1.195}$$

As an example, consider

$$\begin{aligned} \psi &= \mathrm{tr}(\mathbf{A}) = A_{kk}, \\ \psi' &= A'_{kk} = Q_{ki}Q_{kj}A_{ij} = \delta_{ij}A_{ij} = A_{jj} = \psi. \end{aligned} \tag{1.196}$$

ψ and ψ' have the same value and form. If we define ψ and ψ' such that

$$\begin{aligned} \psi &= A_{11}, \\ \psi' &= A_{11} = Q_{j1}Q_{k1}A'_{jk}, \end{aligned} \tag{1.197}$$

then ψ and ψ' have the same value but different functional forms. If we define ψ and ψ' by

$$\begin{aligned} \psi &= A_{11}, \\ \psi' &= A'_{11} = Q_{1j}Q_{1k}A_{jk}, \end{aligned} \tag{1.198}$$

then ψ and ψ' have the same functional form but different values in general unless $\mathbf{A}$ is an isotropic tensor. In this case, ψ is not a scalar in general.

The above rotational invariance can be generalized to vector and tensor functions [5]. For example, A vector $\mathbf{q}$ as a function of one vector $\mathbf{v}$ and one tensor $\mathbf{A}$ may be written as

$$q_k = f_k(v_i, A_{ij}). \tag{1.199}$$

Under $\mathbf{Q}$,

$$q_k = Q_{lk} q'_l. \tag{1.200}$$

f_k is said to be form-invariant under $\mathbf{Q}$ if it satisfies

$$q'_l = f_l(v'_i, A'_{ij}). \tag{1.201}$$

Hence, from Eqs. (1.200) and (1.201),

$$f_k(v_i, A_{ij}) = Q_{lk} f_l(v'_i, A'_{ij}), \tag{1.202}$$

$$f_k(v_i, A_{ij}) = Q_{lk} f_l(Q_{ij} v_j, Q_{im} Q_{jn} A_{mn}), \tag{1.203}$$

$$Q_{pk} f_k(v_i, A_{ij}) = f_p(Q_{ij} v_j, Q_{im} Q_{jn} A_{mn}), \tag{1.204}$$

$$\mathbf{Q}\mathbf{f}(\mathbf{v}, \mathbf{A}) = \mathbf{f}(\mathbf{Q}\mathbf{v}, \mathbf{Q}\mathbf{A}\mathbf{Q}^T). \tag{1.205}$$

A tensor $\mathbf{t}$ as a function of one vector $\mathbf{v}$ and one tensor $\mathbf{A}$ may be written as

$$t_{kl} = f_{kl}(v_i, A_{ij}). \tag{1.206}$$

Under $\mathbf{Q}$,

$$t_{kl} = Q_{pk} Q_{ql} t'_{pq}. \tag{1.207}$$

f_{kl} is said to be form-invariant under $\mathbf{Q}$ if it satisfies

$$t'_{pq} = f_{pq}(v'_i, A'_{ij}). \tag{1.208}$$

Hence,

$$f_{kl}(v_i, A_{ij}) = Q_{pk} Q_{ql} f_{pq}(v'_i, A'_{ij}), \tag{1.209}$$

$$f_{kl}(v_i, A_{ij}) = Q_{pk} Q_{ql} f_{pq}(Q_{ij} v_j, Q_{ir} Q_{js} A_{rs}), \tag{1.210}$$

$$Q_{mk} Q_{nl} f_{kl}(v_i, A_{ij}) = f_{mn}(Q_{ij} v_j, Q_{ir} Q_{js} A_{rs}), \tag{1.211}$$

$$\mathbf{Q}\mathbf{f}(\mathbf{v}, \mathbf{A})\mathbf{Q}^T = \mathbf{f}(\mathbf{Q}\mathbf{v}, \mathbf{Q}\mathbf{A}\mathbf{Q}^T). \tag{1.212}$$

Problems

1.1. Given $A_{ij} = A_{ji}$ and $B_{ij} = -B_{ji}$, show that $A_{ij}B_{ij} = 0$.
1.2. Show that $\varepsilon_{ijk}b_j b_k = 0$.
1.3. Given that $\varepsilon_{ijk}A_{jk} = 0$, show that $A_{jk} = A_{kj}$.
1.4. Show that

$$(P_{ijk} + P_{jki} + P_{jik})x_i x_j x_k = 3P_{ijk}x_i x_j x_k.$$

1.5. Consider an antisymmetric second-order tensor $\mathbf{A} = -\mathbf{A}^T$. Introduce a vector $\mathbf{b}$ through $b_i = \varepsilon_{ijk}A_{kj}/2$. Show that $A_{qp} = \varepsilon_{ipq}b_i$ and that for any vector $\mathbf{c}$, we have $\mathbf{b} \times \mathbf{c} = \mathbf{A} \cdot \mathbf{c}$.
1.6. Given tensors A_{ij} and B_{ij}, show that $\det(\mathbf{A} \cdot \mathbf{B}) = \det(\mathbf{A}) \cdot \det(\mathbf{B})$.
1.7. Show Eq. (1.25) by multiplying Eq. (1.15) by Eq. (1.24) and using Problem 1.6.
1.8. Prove the vector identity

$$(\mathbf{a} \times \mathbf{b}) \cdot (\mathbf{c} \times \mathbf{d}) = (\mathbf{a} \cdot \mathbf{c})(\mathbf{b} \cdot \mathbf{d}) - (\mathbf{a} \cdot \mathbf{d})(\mathbf{b} \cdot \mathbf{c}).$$

1.9. Prove the vector identity

$$\nabla \times (\mathbf{a} \times \mathbf{b}) = (\mathbf{b} \cdot \nabla)\mathbf{a} - \mathbf{b}(\nabla \cdot \mathbf{a}) + \mathbf{a}(\nabla \cdot \mathbf{b}) - (\mathbf{a} \cdot \nabla)\mathbf{b}.$$

1.10. For $F = A_{klX_k X_l}$, show that

$$\frac{\partial F}{\partial x_i} = (A_{ij} + A_{ji})x_j, \quad \frac{\partial F}{\partial x_i \partial x_j} = A_{ij} + A_{ji}.$$

1.11. Prove Green's identity

$$\int_V (\varphi \nabla^2 \psi - \psi \nabla^2 \varphi) dV = \int_S \left(\varphi \frac{\partial \psi}{\partial n} - \psi \frac{\partial \varphi}{\partial n} \right) dS.$$

1.12. Let $\mathbf{A}$, $\mathbf{B}$ and $\mathbf{C}$ be tensors of second or higher orders, show that

$$(\mathbf{A} \cdot \mathbf{B}) \cdot \mathbf{C} = \mathbf{A} \cdot (\mathbf{B} \cdot \mathbf{C}).$$

1.13. Let $\mathbf{A}$ be a second-order tensor and

$$\mathbf{B} = \mathbf{A} \cdot \mathbf{A}^T, \quad \mathbf{C} = \mathbf{A}^T \cdot \mathbf{A}.$$

Show that $\mathbf{B}$ and $\mathbf{C}$ have the same eigenvalues.

1.14. Show that $\text{tr}(\mathbf{A} \cdot \mathbf{A})$ is rotationally invariant.
1.15. Show that $\det(\mathbf{A})$ is rotationally invariant.
1.16. Show that a rotationally invariant scalar function of a three-dimensional vector $\mathbf{v}$ is a function of $\mathbf{v} \cdot \mathbf{v}$.
1.17. Show that the eigenvalues of a second-order tensor $\mathbf{A}$ are rotationally invariant.
1.18. Show that a rotationally invariant scalar function of a second-order, real and symmetric tensor $\mathbf{A}$ is a function of $\text{tr}(\mathbf{A}), \text{tr}(\mathbf{A} \cdot \mathbf{A})$ and $\text{tr}(\mathbf{A} \cdot \mathbf{A} \cdot \mathbf{A})$.
1.19. Show that the components of $\mathbf{v}'$ and $\mathbf{v}$ are related by Eq. (1.175), where $\mathbf{v}'$ is obtained by rotating $\mathbf{v}$ counterclockwise through θ as shown in the figure.

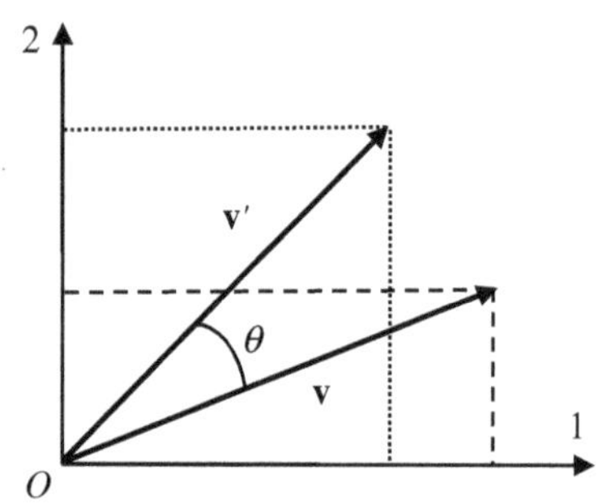

1.20. For a 2×2 matrix, show the Cayley–Hamilton theorem.
1.21. Generalize the Stokes' theorem in Appendix 4 to the case when S has a discontinuity line.
1.22. Show that the A_{ijkl} in Eq. (1.152) is an isotropic tensor.

Chapter 2

Deformation and Motion

In this chapter, we develop the nonlinear kinematics of a continuum with large deformations. The two-point Cartesian tensor notation [5] is used.

2.1 Coordinate Systems

Consider a deformable body which, in the reference configuration at time t_0, occupies a region V with a boundary surface S (see Fig. 2.1). $\mathbf{N}$ is the outward unit normal of S. The position of a material point in this state is denoted by a position vector $\mathbf{X} = X_K \mathbf{I}_K$ in a rectangular coordinate system. X_K are the reference, material or Lagrangian coordinates of the material point. They are continuously varying labels of material particles so that they are identifiable. At time t, the body occupies a region v with a boundary surface s and an outward unit normal $\mathbf{n}$. The current position of the material point associated with $\mathbf{X}$ is given by $\mathbf{y} = y_k \mathbf{i}_k$, which denotes the present, spatial or Eulerian coordinates of the material point. $\mathbf{u}$ is the displacement vector.

The coordinate systems are assumed to be orthogonal, i.e.,

$$\mathbf{i}_k \cdot \mathbf{i}_l = \delta_{kl}, \quad \mathbf{I}_K \cdot \mathbf{I}_L = \delta_{KL}, \tag{2.1}$$

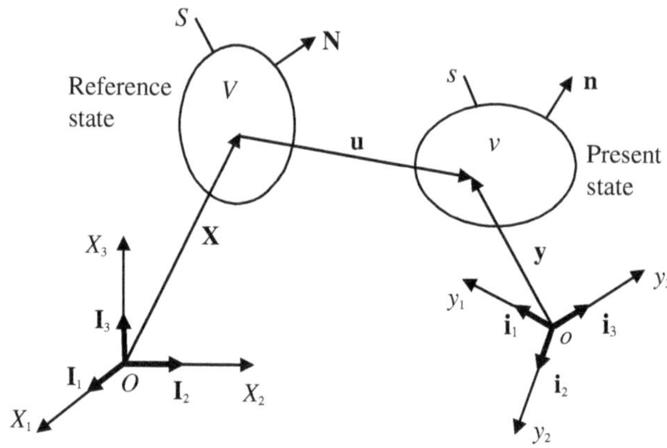

Fig. 2.1. Reference and present states of a continuum and coordinate systems.

where δ_{kl} and δ_{KL} are the Kronecker delta. The transformation coefficients between the basis vectors of the two coordinate systems are denoted by

$$\mathbf{i}_k \cdot \mathbf{I}_L = \delta_{kL} = \delta_{Lk} = \mathbf{I}_L \cdot \mathbf{i}_k. \tag{2.2}$$

In the rest of this book, the two coordinate systems are chosen to be coincident, i.e.,

$$o = O, \quad \mathbf{i}_1 = \mathbf{I}_1, \quad \mathbf{i}_2 = \mathbf{I}_2, \quad \mathbf{i}_3 = \mathbf{I}_3. \tag{2.3}$$

Then δ_{kL} becomes the Kronecker delta. A vector can be resolved into rectangular components in different coordinate systems. For example, we can also write

$$\mathbf{y} = y_K \mathbf{I}_K, \tag{2.4}$$

with

$$y_M = \delta_{Mi} y_i. \tag{2.5}$$

2.2 Motion of a Continuum

The motion of the continuum is described by

$$y_i = y_i(\mathbf{X}, t). \tag{2.6}$$

The trace of a particle $\mathbf{X}$ as t varies is called the path line of the particle. Thus,

$$y_i = y_i(\mathbf{X}, t), \quad \mathbf{X} \text{ fixed} \tag{2.7}$$

gives the path line of the particle initially at $\mathbf{X}$. The displacement vector $\mathbf{u}$ is related to $\mathbf{X}$ and $\mathbf{y}$ through

$$\mathbf{y} = \mathbf{X} + \mathbf{u}, \tag{2.8}$$

or

$$y_i = \delta_{iM}(X_M + u_M). \tag{2.9}$$

Unless otherwise indicated, $\partial/\partial t$ is the time derivative of a field $\phi(\mathbf{y}, t)$ with $\mathbf{y}$ fixed. The material time derivative of a field $\psi(\mathbf{X}, t)$ with $\mathbf{X}$ fixed is denoted by d/dt or a superimposed dot. The velocity and acceleration of the material point associated with $\mathbf{X}$ are given by the following material time derivatives:

$$v_i = \frac{dy_i}{dt} = \dot{y}_i = \left.\frac{\partial y_i(\mathbf{X}, t)}{\partial t}\right|_{\mathbf{X} \text{ fixed}}, \tag{2.10}$$

$$a_i = \frac{dv_i}{dt} = \dot{v}_i = \ddot{y}_i = \left.\frac{\partial^2 y_i(\mathbf{X}, t)}{\partial t^2}\right|_{\mathbf{X} \text{ fixed}}. \tag{2.11}$$

The stream lines at time t are the curves that are tangent to the velocity field $\mathbf{v}(\mathbf{y}, t)$. Hence, the stream lines are determined by

$$\frac{dy_1}{v_1} = \frac{dy_2}{v_2} = \frac{dy_3}{v_3}. \tag{2.12}$$

For a field in the form of $\phi[\mathbf{y}(\mathbf{X}, t), t]$, the material time derivative is given by

$$\begin{aligned} \frac{d\phi}{dt} = \dot{\phi} &= \left.\frac{\partial \phi}{\partial t}\right|_{\mathbf{y} \text{ fixed}} + \left.\frac{\partial \phi}{\partial y_k}\frac{\partial y_k}{\partial t}\right|_{\mathbf{X} \text{ fixed}} \\ &= \frac{\partial \phi}{\partial t} + v_k \frac{\partial \phi}{\partial y_k} = \frac{\partial \phi}{\partial t} + \mathbf{v} \cdot \nabla \phi, \end{aligned} \tag{2.13}$$

where the spatial gradient operator is defined by

$$\nabla = \mathbf{i}_k \frac{\partial}{\partial y_k}. \tag{2.14}$$

2.3 Deformation Gradient

An infinitesimal material line element $\mathbf{dX}$ at t_0 deforms into the following line element at t:

$$dy_i|_{t \text{ fixed}} = y_{i,K}dX_K, \tag{2.15}$$

where the deformation gradient

$$F_{kK} = y_{k,K} = \delta_{kK} + u_{k,K} \tag{2.16}$$

is a two-point tensor in the sense that it depends on both of the two coordinate systems for $\mathbf{y}$ and $\mathbf{X}$. Corresponding to the polar decompositions of matrices in Eq. (1.67), for F_{iK}, we have

$$\begin{aligned} F_{iK} &= R_{iL}U_{LK} = V_{ij}R_{jK}, \\ dy_i &= R_{iL}U_{LK}dX_K = V_{ij}R_{jK}dX_K, \end{aligned} \tag{2.17}$$

where R_{jK} is called the rotation tensor. $\mathbf{U}$ and $\mathbf{V}$ are the right and left stretch tensors. The following determinant is referred to as the Jacobian of the deformation, which is assumed to be positive:

$$\begin{aligned} J &= \det(y_{k,K}) = \varepsilon_{ijk}y_{i,1}y_{j,2}y_{k,3} \\ &= \varepsilon_{KLM}y_{1,K}y_{2,L}y_{3,M} = \frac{1}{6}\varepsilon_{klm}\varepsilon_{KLM}y_{k,K}y_{l,L}y_{m,M} > 0. \end{aligned} \tag{2.18}$$

It can be verified directly that for any L, M and N, the following is true:

$$\varepsilon_{ijk}y_{i,L}y_{j,M}y_{k,N} = J\varepsilon_{LMN}, \tag{2.19}$$

which corresponds to Eq. $(1.72)_1$. From Eq. (2.19), the following can be shown:

$$\begin{aligned} \varepsilon_{ijk}y_{j,M}y_{k,N} &= J\varepsilon_{LMN}X_{L,i}, \\ \varepsilon_{ijk}y_{k,N} &= J\varepsilon_{LMN}X_{L,i}X_{M,j}. \end{aligned} \tag{2.20}$$

Equation $(2.20)_1$ corresponds to Eq. (1.77).

Proof. Multiplying both sides of Eq. (2.19) by $X_{L,r}$, we have

$$\varepsilon_{ijk}y_{i,L}X_{L,r}y_{j,M}y_{k,N} = J\varepsilon_{LMN}X_{L,r}. \tag{2.21}$$

Then

$$\varepsilon_{ijk}\delta_{ir}y_{j,M}y_{k,N} = J\varepsilon_{LMN}X_{L,r}, \tag{2.22}$$

or

$$\varepsilon_{rjk}y_{j,M}y_{k,N} = J\varepsilon_{LMN}X_{L,r}. \tag{2.23}$$

Replacing the index r by i in Eq. (2.23) gives Eq. $(2.20)_1$. Similarly, Eq. $(2.20)_2$ can be proved by multiplying both sides of Eq. $(2.20)_1$ by $X_{M,p}$.

The following relationship can be established:

$$\varepsilon_{ijk}\varepsilon_{LMN}y_{j,M}y_{k,N} = 2JX_{L,i}, \tag{2.24}$$

which corresponds to Eq. (1.79).

Proof. Multiplying both sides of Eq. $(2.20)_1$ by ε_{PMN} gives

$$\begin{aligned} \varepsilon_{ijk}\varepsilon_{PMN}y_{j,M}y_{k,N} &= J\varepsilon_{LMN}\varepsilon_{PMN}X_{L,i} = J\varepsilon_{MNL}\varepsilon_{MNP}X_{L,i} \\ &= J(\delta_{NN}\delta_{LP} - \delta_{NP}\delta_{LN})X_{L,i} = J(3\delta_{LP} - \delta_{LP})X_{L,i} \\ &= 2JX_{P,i}, \end{aligned} \tag{2.25}$$

where the ε–δ identity has been used. Replacing the index P by L in Eq. (2.25) gives Eq. (2.24).

The derivative of the Jacobian with respect to one of its elements is

$$\frac{\partial J}{\partial y_{i,L}} = JX_{L,i}. \tag{2.26}$$

Proof. From Eq. (2.18),

$$\begin{aligned} 6\frac{\partial J}{\partial y_{p,Q}} &= \varepsilon_{ijk}\varepsilon_{LMN}\delta_{ip}\delta_{LQ}y_{j,M}y_{k,N} \\ &\quad + \varepsilon_{ijk}\varepsilon_{LMN}y_{i,L}\delta_{jp}\delta_{MQ}y_{k,N} + \varepsilon_{ijk}\varepsilon_{LMN}y_{i,L}y_{j,M}\delta_{kp}\delta_{NQ} \\ &= \varepsilon_{pjk}\varepsilon_{QMN}y_{j,M}y_{k,N} + \varepsilon_{ipk}\varepsilon_{LQN}y_{i,L}y_{k,N} + \varepsilon_{ijp}\varepsilon_{LMQ}y_{i,L}y_{j,M} \\ &= 3\varepsilon_{pjk}\varepsilon_{QMN}y_{j,M}y_{k,N} = 3(2JX_{Q,p}), \end{aligned} \tag{2.27}$$

where Eq. (2.24) has been used.

With Eq. (2.24), it can also be shown that

$$(JX_{K,k})_{,K} = 0. \tag{2.28}$$

Proof. The differentiation of both sides of Eq. (2.24) with respect to X_L gives

$$2(JX_{L,i})_{,L} = \varepsilon_{ijk}\varepsilon_{LMN}(y_{j,ML}y_{k,N} + y_{j,M}y_{k,NL}) = 0, \tag{2.29}$$

because

$$\varepsilon_{LMN}y_{j,ML} = 0, \quad \varepsilon_{LMN}y_{k,NL} = 0. \tag{2.30}$$

Similarly, the following is true:

$$(J^{-1}y_{k,K})_{,k} = 0. \tag{2.31}$$

2.4 Deformation Tensor and Strain

The lengths of a material line element before and after deformation are given by

$$(dL)^2 = dX_K dX_K = \delta_{KL} dX_K dX_L, \tag{2.32}$$

and

$$\begin{aligned}(dl)^2 &= dy_i dy_i = y_{i,K} dX_K y_{i,L} dX_L \\ &= C_{KL} dX_K dX_L,\end{aligned} \tag{2.33}$$

where C_{KL} is the deformation tensor defined by

$$C_{KL} = y_{k,K}y_{k,L} = C_{LK}, \tag{2.34}$$

and

$$\det(\mathbf{C}) = J^2. \tag{2.35}$$

Similarly, we can also define other deformation tensors [5] such as

$$\begin{aligned}c_{ij} &= X_{K,i}X_{K,j} = c_{ji}, \\ b_{kl} &= \overset{-1}{c}_{kl} = y_{k,M}y_{l,M} = b_{lk}, \\ B_{KL} &= \overset{-1}{C}_{KL} = X_{K,m}X_{L,m} = B_{LK}.\end{aligned} \tag{2.36}$$

It is easy to show that

$$J = (\det \mathbf{C})^{1/2} = (\det \mathbf{c})^{-1/2}. \tag{2.37}$$

From Eqs. (2.32) and (2.33), we have

$$\begin{aligned}(dl)^2 - (dL)^2 &= (C_{KL} - \delta_{KL})dX_K dX_L \\ &= 2E_{KL}dX_K dX_L,\end{aligned} \tag{2.38}$$

where the finite strain tensor E_{KL} is defined by

$$E_{KL} = (C_{KL} - \delta_{KL})/2 = (y_{i,K}y_{i,L} - \delta_{KL})/2 = E_{LK}. \tag{2.39}$$

Similarly, we can define other strain tensors such as e_{ij} by [5]

$$\begin{aligned}e_{ij} &= (\delta_{ij} - c_{ij})/2 = (\delta_{ij} - X_{K,i}X_{K,j})/2 = e_{ji} \\ &= E_{KL}X_{K,i}X_{K,j}.\end{aligned} \tag{2.40}$$

In terms of displacements,

$$E_{KL} = (y_{i,K}y_{i,L} - \delta_{KL})/2 = (u_{K,L} + u_{L,K} + u_{M,K}u_{M,L})/2. \tag{2.41}$$

The unabbreviated form of Eq. (2.41) is

$$\begin{aligned}E_{11} &= u_{1,1} + (u_{1,1}u_{1,1} + u_{2,1}u_{2,1} + u_{3,1}u_{3,1})/2, \\ E_{22} &= u_{2,2} + (u_{1,2}u_{1,2} + u_{2,2}u_{2,2} + u_{3,2}u_{3,2})/2, \\ E_{33} &= u_{3,3} + (u_{1,3}u_{1,3} + u_{2,3}u_{2,3} + u_{3,3}u_{3,3})/2,\end{aligned} \tag{2.42}$$

$$\begin{aligned}E_{23} &= (u_{2,3} + u_{3,2} + u_{1,2}u_{1,3} + u_{2,2}u_{2,3} + u_{3,2}u_{3,3})/2, \\ E_{31} &= (u_{3,1} + u_{1,3} + u_{1,3}u_{1,1} + u_{2,3}u_{2,1} + u_{3,3}u_{3,1})/2, \\ E_{12} &= (u_{1,2} + u_{2,1} + u_{1,1}u_{1,2} + u_{2,1}u_{2,2} + u_{3,1}u_{3,2})/2.\end{aligned} \tag{2.43}$$

Consider the case of small displacement gradients, i.e., under some norm, the displacement gradient with respect to the material coordinates is infinitesimal, e.g.,

$$\|u_{i,K}\| \ll 1, \quad \|u_{i,K}\| = \max|u_{i,K}|. \tag{2.44}$$

We have, approximately,

$$E_{KL} \cong \frac{1}{2}(u_{L,K} + u_{K,L}) = \varepsilon_{KL} = \varepsilon_{LK}, \tag{2.45}$$

where the infinitesimal or linear strain components ε_{KL} are related to the displacements through

$$\varepsilon_{11} = u_{1,1}, \quad \varepsilon_{22} = u_{2,2}, \quad \varepsilon_{33} = u_{3,3}, \tag{2.46}$$

$$2\varepsilon_{12} = u_{1,2} + u_{2,1}, \quad 2\varepsilon_{23} = u_{2.3} + u_{3,2}, \quad 2\varepsilon_{31} = u_{3,1} + u_{1,3}. \tag{2.47}$$

For the six strain components derived from the three displacement components, it may be expected that some relationships among the strain components exist. It can be verified directly that the following equations are satisfied by the linear strain-displacement expressions in Eqs. (2.46) and (2.47):

$$\begin{aligned} 2\varepsilon_{23,23} &= \varepsilon_{22,33} + \varepsilon_{33,22}, \\ 2\varepsilon_{31,31} &= \varepsilon_{33,11} + \varepsilon_{11,33}, \\ 2\varepsilon_{12,12} &= \varepsilon_{11,22} + \varepsilon_{22,11}, \end{aligned} \tag{2.48}$$

$$\begin{aligned} \varepsilon_{11,23} &= (\varepsilon_{31,2} + \varepsilon_{12,3} - \varepsilon_{23,1})_{,1}, \\ \varepsilon_{22,31} &= (\varepsilon_{12,3} + \varepsilon_{23,1} - \varepsilon_{31,2})_{,2}, \\ \varepsilon_{33,12} &= (\varepsilon_{23,1} + \varepsilon_{31,2} - \varepsilon_{12,3})_{,3}, \end{aligned} \tag{2.49}$$

which are the compatibility conditions of the linear strains. Hence, the compatibility conditions are necessary conditions for the six strain components derived from the three displacement components. They are also sufficient in the sense that for six strain components satisfying Eqs. (2.48) and (2.49), there exist three displacement components u_K from which the six strain components are derivable according to

$$\varepsilon_{KL} = \frac{1}{2}(u_{L,K} + u_{K,L}). \tag{2.50}$$

The sufficiency of Eqs. (2.48) and (2.49) can be shown by path-independent line integrals in a singly-connected domain [6, 7]. For a multiply-connected domain, some additional conditions are needed [8]. The six compatibility conditions are not independent. There are three differential relationships among them (Bianchi identity) [2, 9]. The compatibility conditions can also be obtained from the fact that the curvature of a Euclidean space is zero [2]. In a more compact form, Eqs. (2.48) and (2.49) can be written as

$$\varepsilon_{IJ,KL} + \varepsilon_{KL,IJ} - \varepsilon_{IK,JL} - \varepsilon_{JL,IK} = 0. \tag{2.51}$$

The displacement gradient may be decomposed into symmetric and antisymmetric parts as

$$u_{J,I} = \varepsilon_{IJ} + \omega_{IJ}, \tag{2.52}$$

where

$$\omega_{IJ} = \frac{1}{2}(u_{J,I} - u_{I,J}), \tag{2.53}$$

or

$$\begin{aligned}\omega_{23} &= \frac{1}{2}\left(\frac{\partial u_3}{\partial X_2} - \frac{\partial u_2}{\partial X_3}\right) = \omega_1,\\ \omega_{31} &= \frac{1}{2}\left(\frac{\partial u_1}{\partial X_3} - \frac{\partial u_3}{\partial X_1}\right) = \omega_2,\\ \omega_{12} &= \frac{1}{2}\left(\frac{\partial u_2}{\partial X_1} - \frac{\partial u_1}{\partial X_2}\right) = \omega_3,\end{aligned} \tag{2.54}$$

where, corresponding to the antisymmetric ω_{IJ}, we have introduced a vector ω_K. Next, we examine the geometric interpretations of $\boldsymbol{\varepsilon}$ and $\boldsymbol{\omega}$. For simplicity, we introduce the following notation:

$$\begin{aligned}(x, y, z) &= (X_1, X_2, X_3),\\ (u, v, w) &= (u_1, u_2, u_3).\end{aligned} \tag{2.55}$$

Consider two infinitesimal and orthogonal material fibers PM and PN before deformation as shown in Fig. 2.2. After deformation, PM becomes $P'M'$. We have

$$\begin{aligned}(P'M')^2 &= \left(\Delta x + \frac{\partial u}{\partial x}\Delta x\right)^2 + \left(\frac{\partial v}{\partial x}\Delta x\right)^2\\ &\cong (\Delta x)^2 + 2(\Delta x)^2\frac{\partial u}{\partial x},\end{aligned} \tag{2.56}$$

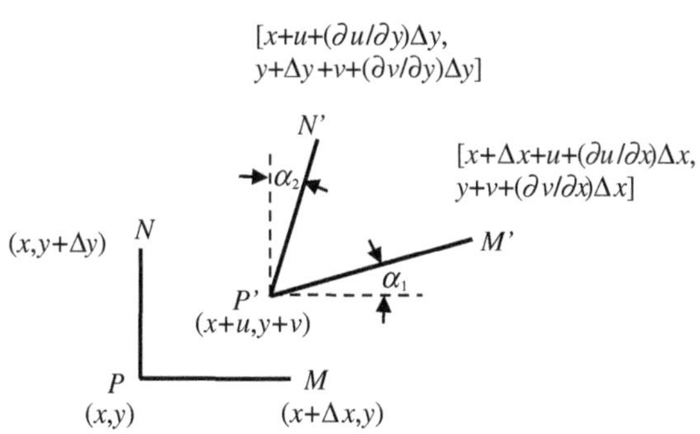

Fig. 2.2. Displacements of two infinitesimal and orthogonal fibers.

where the approximation is based on the small displacement gradient assumption, e.g., $|\partial u/\partial x| \ll 1$. Then,

$$P'M' \cong \Delta x \left(1 + 2\frac{\partial u}{\partial x}\right)^{1/2} \cong \Delta x \left(1 + \frac{\partial u}{\partial x}\right). \tag{2.57}$$

Hence,

$$\varepsilon_{11} = \frac{P'M' - PM}{PM} = \frac{\Delta x\,(1 + \partial u/\partial x) - \Delta x}{\Delta x} = \frac{\partial u}{\partial x}. \tag{2.58}$$

Similarly,

$$\varepsilon_{22} = \frac{\partial v}{\partial y}. \tag{2.59}$$

From Fig. 2.2, we also have

$$\alpha_1 \cong \tan \alpha_1 = \frac{\Delta x \partial v/\partial x}{\Delta x(1 + \partial u/\partial x)} \cong \frac{\partial v}{\partial x}, \tag{2.60}$$

$$\alpha_2 \cong \tan \alpha_2 = \frac{\Delta y \partial u/\partial y}{\Delta y(1 + \partial v/\partial y)} \cong \frac{\partial u}{\partial y}. \tag{2.61}$$

Then the so-called engineering shear strain γ_{12} is defined by

$$\gamma_{12} = 2\varepsilon_{12} = \alpha_1 + \alpha_2 \cong \frac{\partial v}{\partial x} + \frac{\partial u}{\partial y}. \tag{2.62}$$

The average counterclockwise rotation of the two fibers is given by

$$\omega_3 = \frac{\alpha_1 - \alpha_2}{2} \cong \frac{1}{2}\left(\frac{\partial v}{\partial x} - \frac{\partial u}{\partial y}\right). \tag{2.63}$$

2.5 Changes of Area and Volume Elements

At the same material point, consider two material line elements $\mathbf{d}\overset{(1)}{\mathbf{X}}$ and $\mathbf{d}\overset{(2)}{\mathbf{X}}$ which deform into $\mathbf{d}\overset{(1)}{\mathbf{y}}$ and $\mathbf{d}\overset{(2)}{\mathbf{y}}$. The area of the parallelogram spanned by $\mathbf{d}\overset{(1)}{\mathbf{X}}$ and $\mathbf{d}\overset{(2)}{\mathbf{X}}$, and that by $\mathbf{d}\overset{(1)}{\mathbf{y}}$ and $\mathbf{d}\overset{(2)}{\mathbf{y}}$, can be represented by the following vectors, respectively,

$$N_L dS = dS_L = \varepsilon_{LMN} d\overset{(1)}{X}_M \, d\overset{(2)}{X}_N, \tag{2.64}$$

$$n_i ds = ds_i = \varepsilon_{ijk} d\overset{(1)}{y}_j \, d\overset{(2)}{y}_k. \tag{2.65}$$

They are related by

$$ds_i = J X_{L,i} dS_L. \tag{2.66}$$

Proof.

$$\begin{aligned} ds_i &= \varepsilon_{ijk} d\overset{(1)}{y}_j \, d\overset{(2)}{y}_k \\ &= \varepsilon_{ijk} y_{j,M} d\overset{(1)}{X}_M \, y_{k,N} d\overset{(2)}{X}_N = \varepsilon_{ijk} y_{j,M} y_{k,N} d\overset{(1)}{X}_M \, d\overset{(2)}{X}_N \\ &= J X_{L,i} \varepsilon_{LMN} d\overset{(1)}{X}_M \, d\overset{(2)}{X}_N = J X_{L,i} dS_L, \end{aligned} \tag{2.67}$$

where Eq. $(2.20)_1$ has been used.

At the same material point, consider three material line elements $\mathbf{d}\overset{(1)}{\mathbf{X}}$, $\mathbf{d}\overset{(2)}{\mathbf{X}}$ and $\mathbf{d}\overset{(3)}{\mathbf{X}}$ which deform into $\mathbf{d}\overset{(1)}{\mathbf{y}}$, $\mathbf{d}\overset{(2)}{\mathbf{y}}$ and $\mathbf{d}\overset{(3)}{\mathbf{y}}$. The volume of the parallelepiped spanned by $\mathbf{d}\overset{(1)}{\mathbf{X}}$, $\mathbf{d}\overset{(2)}{\mathbf{X}}$ and $\mathbf{d}\overset{(3)}{\mathbf{X}}$, and that by $\mathbf{d}\overset{(1)}{\mathbf{y}}$, $\mathbf{d}\overset{(2)}{\mathbf{y}}$ and $\mathbf{d}\overset{(3)}{\mathbf{y}}$, are related by

$$dv = JdV. \tag{2.68}$$

Proof.

$$\begin{aligned} dv &= \mathbf{d}\overset{(1)}{\mathbf{y}} \cdot (\mathbf{d}\overset{(2)}{\mathbf{y}} \times \mathbf{d}\overset{(3)}{\mathbf{y}}) = \varepsilon_{ijk} d\overset{(1)}{y}_i \, d\overset{(2)}{y}_j \, d\overset{(3)}{y}_k \\ &= \varepsilon_{ijk} y_{i,L} d\overset{(1)}{X}_L \, y_{j,M} d\overset{(2)}{X}_M \, y_{k,N} d\overset{(3)}{X}_N \\ &= \varepsilon_{ijk} y_{i,L} y_{j,M} y_{k,N} d\overset{(1)}{X}_L \, d\overset{(2)}{X}_M \, d\overset{(3)}{X}_N \\ &= J \varepsilon_{LMN} d\overset{(1)}{X}_L \, d\overset{(2)}{X}_M \, d\overset{(3)}{X}_N \\ &= J\mathbf{d}\overset{(1)}{\mathbf{X}} \cdot (\mathbf{d}\overset{(2)}{\mathbf{X}} \times \mathbf{d}\overset{(3)}{\mathbf{X}}) = JdV, \end{aligned} \tag{2.69}$$

where Eq. (2.19) has been used.

2.6 Deformation Rates

The gradient of the velocity field $\mathbf{v}(\mathbf{y}, t)$ is defined by

$$\begin{aligned} \nabla\mathbf{v} &= (\partial_k \mathbf{i}_k)(v_l \mathbf{i}_l) = v_{l,k}\mathbf{i}_k\mathbf{i}_l, \\ (\nabla\mathbf{v})_{kl} &= v_{l,k}. \end{aligned} \tag{2.70}$$

We also denote

$$\begin{aligned} \mathbf{v}\nabla &= (v_k \mathbf{i}_k)(\partial_l \mathbf{i}_l) = v_{k,l}\mathbf{i}_k\mathbf{i}_l, \\ (\mathbf{v}\nabla)_{kl} &= v_{k,l}. \end{aligned} \tag{2.71}$$

The deformation rate tensor d_{ij} and the spin tensor Ω_{ij} are introduced by decomposing the velocity gradient into symmetric and anti-symmetric parts, i.e.,

$$\begin{gathered} v_{j,i} = d_{ij} + \Omega_{ij}, \\ \nabla\mathbf{v} = \mathbf{d} + \boldsymbol{\Omega}, \end{gathered} \tag{2.72}$$

where

$$\begin{aligned} d_{ij} &= \frac{1}{2}(v_{j,i} + v_{i,j}) = d_{ji}, \\ \mathbf{d} &= \frac{1}{2}(\nabla\mathbf{v} + \mathbf{v}\nabla) = \mathbf{d}^T, \end{aligned} \tag{2.73}$$

$$\begin{aligned} \Omega_{ij} &= \frac{1}{2}(v_{j,i} - v_{i,j}) = -\Omega_{ji}, \\ \boldsymbol{\Omega} &= \frac{1}{2}(\nabla\mathbf{v} - \mathbf{v}\nabla) = -\boldsymbol{\Omega}^T. \end{aligned} \tag{2.74}$$

We note that the above definitions of the displacement gradient tensor and the spin tensor differ from those in many other books by a transpose operation. d_{ij} and Ω_{ij} have similar geometric interpretations as ε_{ij} and ω_{ij} but are for deformation rates and rotation rates. The curl of the velocity field, $\nabla \times \mathbf{v}$, is referred to as the vorticity vector. We also have

$$\begin{aligned} \frac{d}{dt}(dy_i) &= \frac{d}{dt}\left(\frac{\partial y_i}{\partial X_K} dX_K\right) = \frac{\partial}{\partial X_K}\left(\frac{dy_i}{dt}\right) dX_K \\ &= \frac{\partial}{\partial X_K}(v_i) dX_K = v_{i,K} dX_K = v_{i,j} y_{j,K} dX_K. \end{aligned} \tag{2.75}$$

The strain rate and the deformation rate are related by

$$\dot{E}_{KL} = d_{ij} y_{i,K} y_{j,L}. \tag{2.76}$$

Proof.

$$\begin{aligned}
\dot{E}_{KL} &= \frac{1}{2}(\dot{y}_{i,K} y_{i,L} + y_{i,K} \dot{y}_{i,L}) = \frac{1}{2}(v_{i,K} y_{i,L} + y_{i,K} v_{i,L}) \\
&= \frac{1}{2}(v_{i,j} y_{j,K} y_{i,L} + y_{i,K} v_{i,j} y_{j,L}) \\
&= \frac{1}{2}(v_{j,i} y_{i,K} y_{j,L} + y_{i,K} v_{i,j} y_{j,L}) \\
&= \frac{1}{2}(v_{j,i} + v_{i,j}) y_{i,K} y_{j,L} = d_{ij} y_{i,K} y_{j,L}.
\end{aligned} \tag{2.77}$$

The material derivative of the Jacobian is given by

$$\dot{J} = J v_{k,k}. \tag{2.78}$$

Proof. From Eq. (2.18),

$$\begin{aligned}
\dot{J} &= \frac{1}{6}\varepsilon_{klm}\varepsilon_{KLM}(v_{k,K} y_{l,L} y_{m,M} + y_{k,K} v_{l,L} y_{m,M} + y_{k,K} y_{l,L} v_{m,M}) \\
&= \frac{1}{2}\varepsilon_{klm}\varepsilon_{KLM} v_{k,K} y_{l,L} y_{m,M} = \frac{1}{2} v_{k,K} \varepsilon_{klm}\varepsilon_{KLM} y_{l,L} y_{m,M} \\
&= \frac{1}{2} v_{k,K} 2 J X_{K,k} = J v_{k,k},
\end{aligned} \tag{2.79}$$

where Eq. (2.24) has been used.

The following expressions are also useful:

$$\begin{aligned}
&\frac{d}{dt}(dv) = v_{k,k} dv, \quad \frac{d}{dt}(dy_k) = v_{k,l} dy_l, \\
&\frac{d}{dt}(y_{k,K}) = v_{k,l} y_{l,K}, \quad \frac{d}{dt}(X_{K,k}) = -v_{l,k} X_{K,l}, \\
&\frac{d}{dt}(dl)^2 = 2 d_{kl} dy_k dy_l, \quad \frac{d}{dt}(ds_k) = v_{m,m} ds_k - v_{m,k} ds_m.
\end{aligned} \tag{2.80}$$

Proof of Eq. $(2.80)_1$:

$$\frac{d}{dt}(dv) = \frac{d}{dt}(J dV) = \frac{dJ}{dt} dV = J v_{k,k} dV = v_{k,k} dv, \tag{2.81}$$

where Eq. (2.78) has been used.

2.7 Material Time Derivatives of Volume Integrals

The material time derivative of a volume integral over a material body is

$$\begin{aligned} \frac{d}{dt}\int_v \phi dv &= \frac{d}{dt}\int_V \phi J dV = \int_V \frac{d}{dt}(\phi J) dV \\ &= \int_v \left(\dot{\phi} + \phi v_{k,k}\right) dv = \int_v \left[\frac{\partial \phi}{\partial t} + (\phi v_k)_{,k}\right] dv, \end{aligned} \tag{2.82}$$

or

$$\frac{d}{dt}\int_v \phi dv = \int_v \frac{\partial \phi}{\partial t} dv + \int_s n_k v_k \phi ds, \tag{2.83}$$

which is usually referred to as Reynold's transport theorem.

2.8 Integrals Over Arbitrarily Changing Volumes

In this section, we discuss the time rate of change of an integral over an arbitrarily changing volume [1] whose boundary does not have to be a material surface. We begin with a surface moving and deforming in general defined by the following equation:

$$F(z_1, z_2, z_3, t) = 0. \tag{2.84}$$

On the surface, we have

$$\begin{aligned} \delta F &= \frac{\partial F}{\partial t}\delta t + (\nabla F)\cdot \delta \mathbf{z} = 0, \\ \nabla F &= \mathbf{e}_i \frac{\partial F}{\partial z_i}. \end{aligned} \tag{2.85}$$

Then

$$\begin{aligned} &\frac{\partial F}{\partial t} + |\nabla F| \frac{\nabla F}{|\nabla F|}\cdot \frac{\delta \mathbf{z}}{\delta t} \\ &\quad = \frac{\partial F}{\partial t} + |\nabla F| \mathbf{n}\cdot \mathbf{v}^s = \frac{\partial F}{\partial t} + |\nabla F| N = 0, \end{aligned} \tag{2.86}$$

where

$$\mathbf{n} = \frac{\nabla F}{|\nabla F|}, \quad \mathbf{v}^s = \frac{\delta \mathbf{z}}{\delta t},$$
$$N = \mathbf{v}^s \cdot \mathbf{n} = -\frac{\partial F}{\partial t} \Big/ |\nabla F|. \tag{2.87}$$

$\mathbf{n}$ is the unit normal of the surface at $\mathbf{z}$. $\mathbf{v}^s = \delta\mathbf{z}/\delta t$ is the velocity of the point of the surface at $\mathbf{z}$. The use of $\delta/\delta t$ is to distinguish it from a material time derivative. N is the normal velocity of the point of the surface at $\mathbf{z}$.

When the surface is in a continuum whose velocity field is $\mathbf{v} = d\mathbf{y}/dt$, we denote the relative velocity of the surface with respect to the continuum at $\mathbf{y}$ in the direction of $\mathbf{n}$ by

$$\begin{aligned}\Theta &= \mathbf{v}^s \cdot \mathbf{n} - \mathbf{v} \cdot \mathbf{n} = N - \mathbf{v} \cdot \mathbf{n} = N - v_k n_k \\ &= -\frac{\partial F}{\partial t} \Big/ |\nabla F| - \mathbf{v} \cdot \frac{\nabla F}{|\nabla F|} \\ &= \left(-\frac{\partial F}{\partial t} - v_k \frac{\partial F}{\partial y_k}\right) \Big/ |\nabla F| = -\frac{dF}{dt} \Big/ |\nabla F|.\end{aligned} \tag{2.88}$$

In the special case when the surface is a material surface consisting of the same particles of the continuum, $dF/dt = 0$, $\mathbf{v}_s = \mathbf{v}$, $N = v_k n_k$ and Θ vanishes.

For the time rate of change of an integral over an arbitrarily changing volume v, to distinguish it from a material time derivative, we write

$$\frac{\delta}{\delta t} \int_{v(t)} \phi dv = \lim_{\Delta t \to 0} \frac{1}{\Delta t} \left[\int_{v+\Delta v} \phi(\mathbf{y}, t+\Delta t) dv - \int_v \phi(\mathbf{y}, t) dv \right], \tag{2.89}$$

which can be further written as

$$\begin{aligned}\frac{\delta}{\delta t} \int_{v(t)} \phi dv = \lim_{\Delta t \to 0} \frac{1}{\Delta t} \bigg\{ &\int_v [\phi(\mathbf{y}, t+\Delta t) - \phi(\mathbf{y}, t)] dv \\ &+ \int_{\Delta v} \phi(\mathbf{y}, t+\Delta t) dv \bigg\},\end{aligned} \tag{2.90}$$

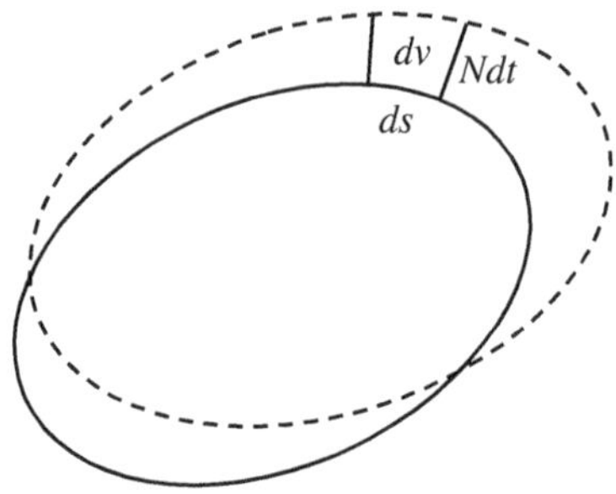

Fig. 2.3. Volume change due to the deformation of a closed surface.

or

$$\frac{\delta}{\delta t}\int_{v(t)}\phi dv=\int_{v}\frac{\partial\phi}{\partial t}dv+\lim_{\Delta t\to 0}\frac{1}{\Delta t}\int_{\Delta v}\phi(\mathbf{y},t+\Delta t)dv. \tag{2.91}$$

From Fig. 2.3,

$$\int_{\Delta v}\phi dv=\int_{s(t)}\phi N\Delta t ds, \tag{2.92}$$

where s is the boundary surface of v. Hence,

$$\frac{\delta}{\delta t}\int_{v(t)}\phi dv=\int_{v(t)}\frac{\partial\phi}{\partial t}dv+\int_{s(t)}\phi N ds. \tag{2.93}$$

In the special case when s is a material surface, $\Theta=0$, $N=v_k n_k$ and Eq. (2.93) reduces to

$$\frac{d}{dt}\int_{v}\phi dv=\int_{v}\frac{\partial\phi}{\partial t}dv+\int_{s}\phi n_k v_k ds, \tag{2.94}$$

which is the transport theorem in Eq. (2.83).

2.9 A Material Body with a Discontinuity Surface

In this section, we discuss the time derivative of an integral over a material body v containing a discontinuity surface Σ as shown in Fig. 2.4 [5]. Physically, Σ is a thin layer in which certain fields change rapidly across its thickness. Mathematically, Σ is treated as a surface across which certain fields are discontinuous and are

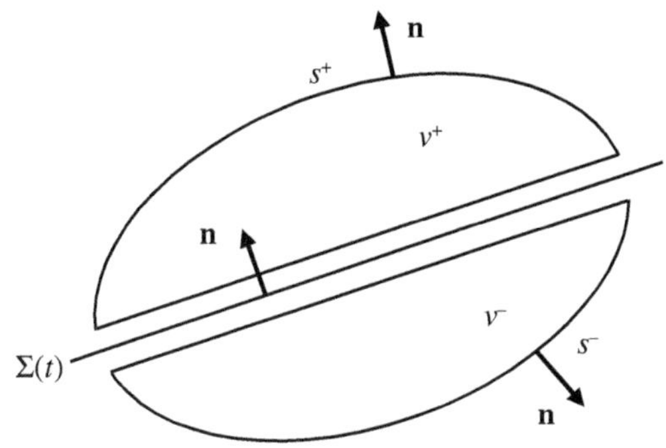

Fig. 2.4. A material body with a discontinuity surface.

not differentiable. Σ does not have to be material or planar. It divides the material body v into two parts. Although each part by itself is not a material body, the two parts together is. We denote

$$v = v^+ \cup v^-, \quad \partial v = s^+ \cup s^- = s, \\ \partial v^+ = s^+ \cup \Sigma, \quad \partial v^- = s^- \cup \Sigma. \tag{2.95}$$

Consider a field ϕ over v which may have a finite jump discontinuity over Σ and its derivative across Σ may be singular, like a delta function. We apply the general time derivative of a volume integral in Eq. (2.93) to the integration of ϕ over each part of v:

$$\frac{\delta}{\delta t}\int_{v^-}\phi dv = \int_{v^-}\frac{\partial \phi}{\partial t}dv + \int_{s^-}\phi v_k n_k ds + \int_{\Sigma}\phi^- N ds, \tag{2.96}$$

$$\frac{\delta}{\delta t}\int_{v^+}\phi dv = \int_{v^+}\frac{\partial \phi}{\partial t}dv + \int_{s^+}\phi v_k n_k ds - \int_{\Sigma}\phi^+ N ds. \tag{2.97}$$

Adding Eqs. (2.96) and (2.97), we obtain the time derivative of the integration of ϕ over the material body containing Σ as

$$\frac{d}{dt}\int_{v}\phi dv = \int_{v-\Sigma}\frac{\partial \phi}{\partial t}dv + \int_{s}\phi v_k n_k ds - \int_{\Sigma}[\phi] N ds, \tag{2.98}$$

where

$$[\phi] = \phi^+ - \phi^-, \tag{2.99}$$

and

$$\int_{v-\Sigma} (\,)dv = \int_{v_1} (\,)dv + \int_{v_2} (\,)dv. \tag{2.100}$$

The integration over $v-\Sigma$ is simply the sum of the integrations over v_1 and v_2. It does not have the contribution from Σ. In the special case when ϕ is continuous across Σ, we have $[\phi] = 0$ and Eq. (2.98) reduces to the transport theorem in Eq. (2.94) or (2.83).

For a tensor field $\mathbf{\Phi}$ over v containing a discontinuity surface Σ, from the divergence theorem in Eq. (1.149), we have

$$\int_s \mathbf{n} \cdot \mathbf{\Phi} ds = \int_{v-\Sigma} \nabla \cdot \mathbf{\Phi} dv + \int_\Sigma \mathbf{n} \cdot [\mathbf{\Phi}] ds. \tag{2.101}$$

As a specific case, we let $\mathbf{\Phi} = \phi \mathbf{v}$ in Eq. (2.101). This yields

$$\int_s \phi v_k n_k ds = \int_{v-\Sigma} (\phi v_k)_{,k} dv + \int_\Sigma [\phi v_k n_k] ds. \tag{2.102}$$

Substituting Eq. (2.102) into Eq. (2.98), we obtain

$$\frac{d}{dt}\int_v \phi dv = \int_{v-\Sigma} \left[\frac{\partial \phi}{\partial t} + (\phi v_k)_{,k}\right] dv + \int_\Sigma [\phi(v_k n_k - N)] ds, \tag{2.103}$$

or

$$\frac{d}{dt}\int_v \phi dv = \int_{v-\Sigma} \left[\frac{\partial \phi}{\partial t} + (\phi v_k)_{,k}\right] dv - \int_\Sigma [\phi \Theta] ds, \tag{2.104}$$

where, from Eq. (2.88),

$$\Theta = N - \mathbf{v} \cdot \mathbf{n}. \tag{2.105}$$

When $[\phi\Theta] = 0$, Eq. (2.104) reduces to Eq. (2.82).

2.10 Objective Tensors

Consider a change of observer from $(\mathbf{y}, t)$ to $(\mathbf{y}', t')$:

$$y'_i = Q_{ij}(t)y_j + c_i(t) \quad \text{or} \quad \mathbf{y}' = \mathbf{Q}(t)\mathbf{y} + \mathbf{c}(t),$$
$$t' = t - \tau, \tag{2.106}$$

where the operations involving the boldface $\mathbf{Q}$ are based on the matrix notation in Section 1.3 for the components of the relevant vectors and tensors, rather than the tensor notation and operation. The same is true for the rest of this section. τ is a constant. $\det(\mathbf{Q}) = 1$. $\mathbf{Q}$ is an orthogonal tensor representing a time-dependent rotation. $\mathbf{c}$ is a vector representing a time-dependent translation. In the special case when $\mathbf{Q}$ does not depend on time and $\mathbf{c}$ is a linear function of time, Eq. (2.106) reduces to the Galilean transformation of a constant rigid rotation, a constant velocity translation and a constant shift of time. Equation $(2.106)_1$ can be inverted as

$$\begin{aligned} \mathbf{y} &= \mathbf{Q}^T(\mathbf{y}' - \mathbf{c}), \\ y_k &= Q_{mk}(y'_m - c_m), \\ \frac{\partial y_k}{\partial y'_m} &= Q_{mk}. \end{aligned} \tag{2.107}$$

Since

$$\mathbf{Q}\mathbf{Q}^T = \mathbf{Q}^T\mathbf{Q} = \mathbf{1}, \tag{2.108}$$

we have

$$\begin{aligned} \dot{\mathbf{Q}}\mathbf{Q}^T + \mathbf{Q}\dot{\mathbf{Q}}^T &= \mathbf{0}, \\ \dot{\mathbf{Q}}^T\mathbf{Q} + \mathbf{Q}^T\dot{\mathbf{Q}} &= \mathbf{0}. \end{aligned} \tag{2.109}$$

Equation $(2.109)_1$ shows that $\dot{\mathbf{Q}}\mathbf{Q}^T$ (and $\mathbf{Q}\dot{\mathbf{Q}}^T$ too) is antisymmetric and may be associated with a vector. Under Eq. (2.106), if a scalar ψ, the components of a vector $\mathbf{q}$ and the components of a tensor $\mathbf{A}$ change according to

$$\begin{aligned} &\psi' = \psi, \\ &q_i' = Q_{ij}q_j, \quad \mathbf{q}' = \mathbf{Q}\mathbf{q}, \\ &A_{ij}' = Q_{ik}Q_{jl}A_{kl}, \quad \mathbf{A}' = \mathbf{Q}\mathbf{A}\mathbf{Q}^T, \end{aligned} \tag{2.110}$$

they are said to be objective [10]. They will be needed in the construction of constitutive relations in later chapters.

Consider the motion of a continuum which is seen as

$$\begin{aligned} \mathbf{y} &= \mathbf{y}(\mathbf{X}, t), \\ \mathbf{y}' &= \mathbf{y}'(\mathbf{X}, t'), \end{aligned} \tag{2.111}$$

by the two observers related by Eq. (2.106). Obviously, the components of $\mathbf{y}$ and $\mathbf{y}'$ are related by

$$\begin{aligned} &y_i' = Q_{ik}y_k + c_i, \\ &\mathbf{y}'(\mathbf{X}, t') = \mathbf{Q}(t)\mathbf{y}(\mathbf{X}, t) + \mathbf{c}(t). \end{aligned} \tag{2.112}$$

As an example, the deformation gradient is objective because

$$F_{iK}' = \left.\frac{\partial y_i'}{\partial X_K}\right|_{t'} = y_{i,K}' = Q_{ij}y_{j,K} = Q_{ij}F_{jK}. \tag{2.113}$$

As another example, the velocity vector is not objective because

$$\begin{aligned} v_i' &= \left.\frac{\partial y_i'}{\partial t'}\right|_{\mathbf{X}} = \dot{Q}_{ik}y_k + Q_{ik}\left.\frac{\partial y_k}{\partial t}\right|_{\mathbf{X}} + \dot{c}_i \\ &= Q_{ik}v_k + \dot{Q}_{ik}y_k + \dot{c}_i. \end{aligned} \tag{2.114}$$

Then

$$\begin{aligned} v_{i,j}' &= \frac{\partial v_i'}{\partial y_j'} = Q_{ik}\frac{\partial v_k}{\partial y_l}\frac{\partial y_l}{\partial y_j'} + \dot{Q}_{ik}\frac{\partial y_k}{\partial y_j'} \\ &= Q_{ik}Q_{jl}v_{k,l} + \dot{Q}_{ik}Q_{jk}, \end{aligned} \tag{2.115}$$

where Eq. $(2.107)_3$ has been used. Equation (2.115) shows that the velocity gradient is not objective. Switching indices i and j in

Eq. (2.115), we obtain

$$\begin{aligned} v'_{j,i} &= Q_{jk}Q_{il}v_{k,l} + \dot{Q}_{jk}Q_{ik}, \\ (\nabla \mathbf{v})' &= \mathbf{Q}(\nabla \mathbf{v})\mathbf{Q}^T + \mathbf{Q}\dot{\mathbf{Q}}^T. \end{aligned} \tag{2.116}$$

If we add Eqs. (2.115) and (2.116), we get

$$2d'_{ij} = v'_{j,i} + v'_{i,j} = Q_{ik}Q_{jl}(v_{l,k} + v_{k,l}) = Q_{ik}Q_{jl}2d_{kl}, \tag{2.117}$$

where Eq. $(2.109)_1$ has been used. Equation (2.117) shows that the deformation rate tensor $\mathbf{d}$ in Eq. (2.73) is objective. The spin tensor $\boldsymbol{\Omega}$ in Eq. (2.74) is obtained by subtraction Eq. (2.115) from Eq. (2.116) which gives

$$\begin{aligned} 2\Omega'_{ij} &= v'_{j,i} - v'_{i,j} = Q_{il}Q_{jk}(v_{k,l} - v_{l,k}) + \dot{Q}_{jk}Q_{ik} - \dot{Q}_{ik}Q_{jk} \\ &= Q_{il}Q_{jk}2\Omega_{lk} + 2\dot{Q}_{jk}Q_{ik}, \end{aligned} \tag{2.118}$$

or

$$\begin{aligned} \Omega'_{ij} &= Q_{ik}Q_{jl}\Omega_{kl} + Q_{ik}\dot{Q}_{jk}, \\ \boldsymbol{\Omega}' &= \mathbf{Q}\boldsymbol{\Omega}\mathbf{Q}^T + \mathbf{Q}\dot{\mathbf{Q}}^T, \end{aligned} \tag{2.119}$$

which shows that the spin tensor is not objective. The multiplication of Eq. (2.119) by Q_{im} gives

$$\dot{Q}_{jm} = Q_{im}\Omega'_{ij} - Q_{jl}\Omega_{ml}, \tag{2.120}$$

which will be useful later. In addition, it can be shown that the following fields are objective:

$$R_{iK}, \quad U_{KL}, \quad V_{ij}, \quad C_{KL}, \quad E_{KL}, \tag{2.121}$$

and that the acceleration vector is not objective.

Let $\mathbf{q}$ be an objective vector field which satisfies

$$\begin{aligned} q'_j(\mathbf{X}, t') &= Q_{jm}(t)q_m(\mathbf{X}, t), \\ \mathbf{q}'(\mathbf{X}, t') &= \mathbf{Q}(t)\mathbf{q}(\mathbf{X}, t). \end{aligned} \tag{2.122}$$

Its material time derivative satisfies

$$\left.\frac{\partial q_j'}{\partial t'}\right|_{\mathbf{X}} = \dot{Q}_{jm} q_m + Q_{jm} \left.\frac{\partial q_m}{\partial t}\right|_{\mathbf{X}}, \tag{2.123}$$

which is not objective. The substitution of Eq. (2.120) into Eq. (2.123) leads to

$$\begin{aligned} \left.\frac{\partial q_j'}{\partial t'}\right|_{\mathbf{X}} - q_i'\Omega_{ij}' &= Q_{jm}\left(\left.\frac{\partial q_m}{\partial t}\right|_{\mathbf{X}} - q_l \Omega_{lm}\right), \\ \frac{d\mathbf{q}'}{dt'} - \mathbf{q}'\boldsymbol{\Omega}' &= \mathbf{Q}\left(\frac{d\mathbf{q}}{dt} - \mathbf{q}\boldsymbol{\Omega}\right), \end{aligned} \tag{2.124}$$

which shows that

$$\frac{dq_j}{dt} - q_i \Omega_{ij} \quad \text{or} \quad \frac{d\mathbf{q}}{dt} - \mathbf{q}\boldsymbol{\Omega} \tag{2.125}$$

is objective. The material time derivative of an objective tensor field has similar behaviors. For example, for a second-order objective tensor $\mathbf{T}$,

$$\begin{aligned} T_{jk}' &= Q_{jm} Q_{kn} T_{mn}, \\ \mathbf{T}' &= \mathbf{Q}\mathbf{T}\mathbf{Q}^T. \end{aligned} \tag{2.126}$$

For its material time derivative,

$$\dot{\overline{T_{jk}'}} = \dot{Q}_{jm} Q_{kn} T_{mn} + Q_{jm} \dot{Q}_{kn} T_{mn} + Q_{jm} Q_{kn} \dot{T}_{mn}, \tag{2.127}$$

which is not objective. With the use of Eq. (2.120), we can write Eq. (2.127) as

$$\dot{\overline{T_{jk}'}} - \Omega_{ij}' T_{ik}' - \Omega_{ik}' T_{ji}' = Q_{jl} Q_{kn}\left(\dot{T}_{ln} - \Omega_{mn} T_{lm} - \Omega_{ml} T_{mn}\right), \tag{2.128}$$

which shows that

$$\dot{T}_{ij} - \Omega_{kj} T_{ik} - \Omega_{ki} T_{kj} \quad \text{or} \quad \dot{\mathbf{T}} - \mathbf{T}\boldsymbol{\Omega} + \boldsymbol{\Omega}\mathbf{T} \tag{2.129}$$

is an objective time rate, called the Jaumann rate.

We summarize the transformation behaviors of some common tensors under Eq. (2.106) [11]:

	Obj.?	Type	Transformation
Jacobian J	Y	Scalar	$J' = J$
Deformation gradient F_{ik}	Y	2-point	$F'_{iK} = Q_{ij}F_{jK}$
Rotation R_{ik}	Y	2-point	$R'_{iK} = Q_{ij}R_{jK}$
Deformation tensor $\mathbf{C}$	Y	Material	$\mathbf{C}' = \mathbf{C}$
Strain $\mathbf{E}$	Y	Material	$\mathbf{E}' = \mathbf{E}$
Strain rate $\dot{\mathbf{E}}$	Y	Material	$\dot{\overline{\mathbf{E}'}} = \dot{\mathbf{E}}$
Right stretch $\mathbf{U}$	Y	Material	$\mathbf{U}' = \mathbf{U}$
Deformation tensor $\mathbf{b}$	Y	Spatial	$\mathbf{b}' = \mathbf{Q}\mathbf{b}\mathbf{Q}^T$
Strain $\mathbf{e}$	Y	Spatial	$\mathbf{e}' = \mathbf{Q}\mathbf{e}\mathbf{Q}^T$
Left stretch $\mathbf{V}$	Y	Spatial	$\mathbf{V}' = \mathbf{Q}\mathbf{V}\mathbf{Q}^T$
Velocity gradient $\nabla\mathbf{v}$	N	Spatial	$(\nabla\mathbf{v})' = \mathbf{Q}(\nabla\mathbf{v})\mathbf{Q}^T + \mathbf{Q}\dot{\mathbf{Q}}^T$
Deformation rate $\mathbf{d}$	Y	Spatial	$\mathbf{d}' = \mathbf{Q}\mathbf{d}\mathbf{Q}^T$
Spin $\boldsymbol{\Omega}$	N	Spatial	$\boldsymbol{\Omega}' = \mathbf{Q}\boldsymbol{\Omega}\mathbf{Q}^T + \mathbf{Q}\dot{\mathbf{Q}}^T$

Problems

2.1. Show that the following displacement field for a homogeneous deformation transforms straight lines into straight lines and planes into planes:

$$u_K = A_{KL}X_L.$$

2.2. The motion of a continuum is given by

$$y_1 = X_1 + \lambda t X_2, \quad y_2 = X_2, \quad y_3 = X_3,$$

where λ is a constant. Calculate u_k, F_{kL}, J, C_{KL}, E_{KL}, v_k, a_k and $\dot{E}_{LM}$.

2.3. Show that the inverses of the deformation tensors $\mathbf{C}$ and $\mathbf{c}$ are given by

$$\overset{-1}{C}_{KL} = X_{K,k}X_{L,k}, \quad \overset{-1}{c}_{kl} = y_{k,K}y_{l,K}.$$

2.4. From $\dot{E}_{LM} = y_{i,L} y_{j,M} d_{ij}$, show that $d_{ij} = X_{L,i} X_{M,j} \dot{E}_{LM}$.
2.5. Show that the acceleration vector, $\mathbf{a}$, may be expressed as

$$\mathbf{a} = \frac{\partial \mathbf{v}}{\partial t} + (\nabla \times \mathbf{v}) \times \mathbf{v} + \frac{1}{2} \nabla (\mathbf{v} \cdot \mathbf{v}).$$

2.6. Show Eq. (2.39).
2.7. Show Eq. (2.40).
2.8. Show that Eq. (2.50) satisfies Eq. (2.51).
2.9. For the following infinitesimal strain field, determine the restrictions on the parameters A_0, A_1, B_0, B_1, C_0, C_1 and C_2 by the compatibility conditions:

$$\begin{aligned}
\varepsilon_{11} &= A_0 + A_1(X_1^2 + X_2^2) + X_1^4 + X_2^4 \\
\varepsilon_{22} &= B_0 + B_1(X_1^2 + X_2^2) + X_1^4 + X_2^4, \\
2\varepsilon_{12} &= C_0 + C_1 xy(X_1^2 + X_2^2 + C_2), \\
\varepsilon_{33} &= \varepsilon_{23} = \varepsilon_{31} = 0.
\end{aligned}$$

2.10. Study the kinematic interpretations of the components of d_{kl} and Ω_{kl}.
2.11. Determine the stream lines of the velocity field

$$v_1 = cy_2, \quad v_2 = v_3 = 0.$$

2.12. Show that for the velocity field

$$v_1 = y_1^2 y_2 + y_2^3, \quad v_2 = -y_1 y_2^2 - y_1^3, \quad v_3 = 0,$$

the stream lines are circular.
2.13. Show that the velocity field $v_i = \varepsilon_{ijk} b_j y_k + c_i$ where b_i and c_i are constant vectors represents a rigid body rotation and determine the vorticity vector $\nabla \times \mathbf{v}$ for this motion.
2.14. Study the special case of Reynold's transport theorem in Eq. (2.82) when $\phi = 1$.
2.15. Show that the material time derivative of a line integral is given by

$$\frac{d}{dt} \int_c \phi dy_k = \int_c \left(\dot{\phi} dy_k + \phi v_{k,l} dy_l \right),$$

where ϕ may be a tensor field of any order.

2.16. Show that the material time derivative of a surface integral is given by

$$\frac{d}{dt}\int_s \phi ds_k = \int_s \left(\dot{\phi} ds_k + \phi v_{l,l} ds_k - \phi v_{l,k} ds_l\right),$$

and, in the case when $\phi = q_k$,

$$\frac{d}{dt}\int_s q_k ds_k = \int_s \left(\frac{\partial q_k}{\partial t} + v_l q_{k,l} + q_k v_{l,l} - q_l v_{k,l}\right) ds_k = \int_s q_k^* ds_k,$$

$$q_k^* = \dot{q}_k + q_k v_{l,l} - q_l v_{k,l}, \quad \mathbf{q}^* = \dot{\mathbf{q}} + \mathbf{q}(\nabla \cdot \mathbf{v}) - (\mathbf{q} \cdot \nabla)\mathbf{v}.$$

Study the case when s has a discontinuity line.

2.17. A spherical surface with a time-dependent radius is given by

$$F(z_1, z_2, z_3, t) = z_1^2 + z_2^2 + z_3^2 - (R + ct)^2 = 0.$$

Calculate $\nabla F, \mathbf{n}, \mathbf{v}^s$ and N.

2.18. Show that the fields in Eq. (2.121) are objective.

2.19. Show that the acceleration vector is not objective.

2.20. Prove that the spatial gradient of an objective scalar field is an objective vector field; the spatial divergence of an objective vector field is an objective scalar field; and the spatial divergence of an objective tensor field is an objective vector field.

2.21. Consider the two vectors, $\mathbf{r}'(t)$ and $\mathbf{r}$, which are related by $\mathbf{r}'(t) = \mathbf{Q}(t) \cdot \mathbf{r}$ in the tensor notation, where $\mathbf{r}$ does not depend on time, and $\mathbf{Q}$ satisfies Eq. (2.108). Show that $\dot{\mathbf{r}}' = \mathbf{A} \cdot \mathbf{r}'$ where $\mathbf{A} = \dot{\mathbf{Q}} \cdot \mathbf{Q}^T$ and $\mathbf{A} = -\mathbf{A}^T$. If $b_i = \varepsilon_{ijk} A_{kj}/2$, then $\dot{\mathbf{r}}' = \mathbf{b} \times \mathbf{r}'$. Hence, $\mathbf{b}$ may be interpreted as an angular velocity vector.

2.22. For the polar decomposition of the deformation gradient tensor, show that $\mathbf{C} = \mathbf{U}^2$ and $\overset{-1}{\mathbf{c}} = \mathbf{V}^2$. Obtain an expression of $\mathbf{R}$.

2.23. Prove Eq. $(2.80)_{2-6}$.

2.24. Given the deformation gradient tensor

$$[F_{iK}] = \begin{bmatrix} 2 & -2 & 0 \\ 1 & 1 & 0 \\ 0 & 0 & 1 \end{bmatrix},$$

calculate $\mathbf{C}, \mathbf{U}, \mathbf{V}$ and $\mathbf{R}$.

Chapter 3

Balance Laws

In this chapter, we postulate the basic physical laws of continuum mechanics and study their implications. The five laws are the conservation of mass, the linear and angular momentum equations, the energy equation and the entropy inequality. We begin with the integral forms of these laws and then localize them for their differential forms and jump conditions across a discontinuity surface.

3.1 Integral Balance Laws

For a material body occupying a region v with a boundary surface s at time t (see Fig. 2.1), the conservation of mass takes the following form:

$$\frac{d}{dt}\int_v \rho dv = 0, \tag{3.1}$$

where ρ is the present mass density. For a continuum under body force $\mathbf{f}$ per unit mass and surface traction $\mathbf{t}$ per unit area, the linear momentum equation is

$$\frac{d}{dt}\int_v \rho\mathbf{v}dv = \int_v \rho\mathbf{f}dv + \int_s \mathbf{t}ds, \tag{3.2}$$

where $\mathbf{v}$ is the velocity field. For a continuum, the angular momentum equation about $\mathbf{y} = 0$ can be written as

$$\frac{d}{dt}\int_v \mathbf{y}\times\rho\mathbf{v}dv = \int_v \mathbf{y}\times\rho\mathbf{f}dv + \int_s \mathbf{y}\times\mathbf{t}ds, \tag{3.3}$$

where $\mathbf{y}$ is the position vector. Let the internal energy density per unit mass be ε, the heat flux vector be $\mathbf{q}$ and the body heat source per unit mass be r. The conservation of energy takes the following form:

$$\frac{d}{dt}\int_v \rho\left(\frac{1}{2}\mathbf{v}\cdot\mathbf{v}+\varepsilon\right)dv$$
$$=\int_v (\rho\mathbf{f}\cdot\mathbf{v}+\rho r)dv+\int_s (\mathbf{t}\cdot\mathbf{v}-\mathbf{n}\cdot\mathbf{q})ds. \qquad (3.4)$$

The second law of thermodynamics is given by

$$\frac{d}{dt}\int_v \rho\eta dv \geq \int_v \frac{\rho r}{\theta}dv - \int_s \frac{\mathbf{q}\cdot\mathbf{n}}{\theta}ds, \qquad (3.5)$$

where η is the entropy per unit mass and θ is the absolute temperature.

3.2 Conservation of Mass

Equation (3.1) states that the total mass of a material body is a constant, which is the total mass in the reference state (see Fig. 2.1), i.e.,

$$\int_v \rho dv = \int_V \rho^0 dV, \qquad (3.6)$$

where ρ^0 is the mass density in the reference state. For the left-hand side of Eq. (3.6), we apply the change of integration variables to the reference state using Eq. (2.6). Then the conservation of mass in Eq. (3.6) takes the following form:

$$\int_V (\rho J - \rho^0)dV = 0. \qquad (3.7)$$

Equation (3.7) holds for V or any part of it. Assuming a continuous integrand, we conclude from Eq. (3.7) that

$$\rho^0 = \rho J. \qquad (3.8)$$

Alternatively, since the total mass is conserved, we have

$$\begin{aligned}\frac{d}{dt}\int_v \rho dv &= \frac{d}{dt}\int_V \rho J dV = \int_V \frac{d}{dt}(\rho J) dV \\ &= \int_V (\dot{\rho}J + \rho\dot{J})dV = \int_V (\dot{\rho}J + \rho J v_{i,i})dV \\ &= \int_v (\dot{\rho} + \rho v_{i,i})dv = 0,\end{aligned} \tag{3.9}$$

where Eq. (2.78) has been used. Then

$$\begin{aligned}&\dot{\rho} + \rho v_{i,i} = 0, \\ &\frac{\partial \rho}{\partial t} + (\rho v_i)_{,i} = 0.\end{aligned} \tag{3.10}$$

From Eqs. (2.80) and (3.10), we obtain

$$\begin{aligned}\frac{d}{dt}(\rho dv) &= \dot{\rho} dv + \rho \frac{d}{dt}(dv) \\ &= -\rho v_{i,i} dv + \rho v_{i,i} dv = 0.\end{aligned} \tag{3.11}$$

Hence, for a differential material element of the body, the conservation of mass may be written as

$$\rho dv = \rho^0 dV. \tag{3.12}$$

With Eqs. (2.78) and (3.10), it can be shown that

$$\frac{d}{dt}\int_v \rho\phi dv = \int_v \rho \frac{d\phi}{dt} dv, \tag{3.13}$$

where ϕ represents a tensor field.

Proof. With the change of integration variables to the reference state, we have

$$\begin{aligned}\frac{d}{dt}\int_v \rho\phi dv &= \frac{d}{dt}\int_V \rho\phi J dV = \int_V \frac{d}{dt}(\rho\phi J)\, dV \\ &= \int_V \left(\dot{\rho}\phi J + \rho\frac{d\phi}{dt}J + \rho\phi\dot{J}\right) dV \\ &= \int_V \left(-\rho v_{i,i}\phi J + \rho\frac{d\phi}{dt}J + \rho\phi J v_{i,i}\right) dV \\ &= \int_V \rho\frac{d\phi}{dt} J dV = \int_v \rho\frac{d\phi}{dt} dv.\end{aligned} \tag{3.14}$$

3.3 Traction Vector

Consider the continuum in Fig. 3.1. It is under the action of some external forces. The body is cut into two halves, which reveals the interaction between the two halves. The interaction is assumed to be a distributed force over the interface between the two halves. Figure 3.1 shows the interaction over a differential area of the interface on the right half of the body only. Over a small area ΔA of the interface at $\mathbf{y}$, we introduce the traction (or stress) vector $\mathbf{t}$ on ΔA with an outward unit normal $\mathbf{n}$ by

$$\mathbf{t}(\mathbf{y}, \mathbf{n}) = \lim_{\Delta A \to 0} \frac{\Delta \mathbf{F}}{\Delta A}. \tag{3.15}$$

Consider the small and very thin disk taken at $\mathbf{y}$ with unit normal vectors $\mathbf{n}$ and $-\mathbf{n}$ on its two faces as shown in Fig. 3.2. The thickness of the disk is essentially negligible. The volume of the disk is a higher-order infinitesimal when compared to its top and bottom faces because of the very thin thickness. We apply the linear momentum equation in Eq. (3.2) to the disk:

$$\mathbf{t}(\mathbf{n})\Delta A + \mathbf{t}(-\mathbf{n})\Delta A \cong 0, \tag{3.16}$$

where, for simplicity, we have dropped the $\mathbf{y}$ in $\mathbf{t}$. The volume integrals in Eq. (3.2) are neglected as higher-order infinitesimals. The

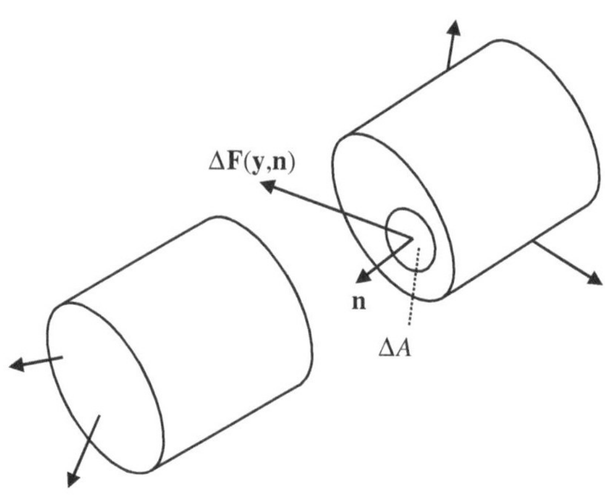

Fig. 3.1. Internal force in a continuum.

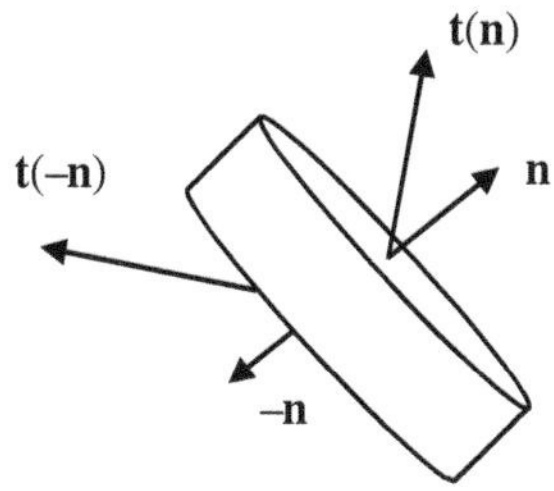

Fig. 3.2. Forces on a small and very thin disk.

contribution from the lateral surface is also neglected for the same reason. Equation (3.16) leads to

$$\mathbf{t}(-\mathbf{n}) = -\mathbf{t}(\mathbf{n}). \tag{3.17}$$

3.4 Stress Tensor

Consider the tetrahedron in Fig. 3.3. Let the areas of the three faces of the tetrahedron with their outward normal directions along $-\mathbf{e}_i$ be dA_i, and the area of the inclined face with a normal $\mathbf{n}$ be dA. In addition to the surface tractions shown, the tetrahedron is also under distributed body forces, which are not shown because they are associated with the volume of the tetrahedron, which is a higher-order infinitesimal when compared to the surface area of the tetrahedron.

For a closed surface s enclosing a volume v, we have the following divergence theorem from Eq. (1.136):

$$\int_s \mathbf{n} \cdot \mathbf{A} ds = \int_v \nabla \cdot \mathbf{A} dv. \tag{3.18}$$

Let

$$\mathbf{A} = \boldsymbol{\delta}. \tag{3.19}$$

Then Eq. (3.18) reduces to

$$\int_s \mathbf{n} ds = 0. \tag{3.20}$$

Specifically, for the tetrahedron in Fig. 3.3, Eq. (3.20) takes the following form:

$$\begin{aligned} &\mathbf{n}dA + (-\mathbf{e}_1)dA_1 + (-\mathbf{e}_2)dA_2 + (-\mathbf{e}_3)dA_3 \\ &= \mathbf{n}dA - \mathbf{e}_1 dA_1 - \mathbf{e}_2 dA_2 - \mathbf{e}_3 dA_3 = 0, \end{aligned} \tag{3.21}$$

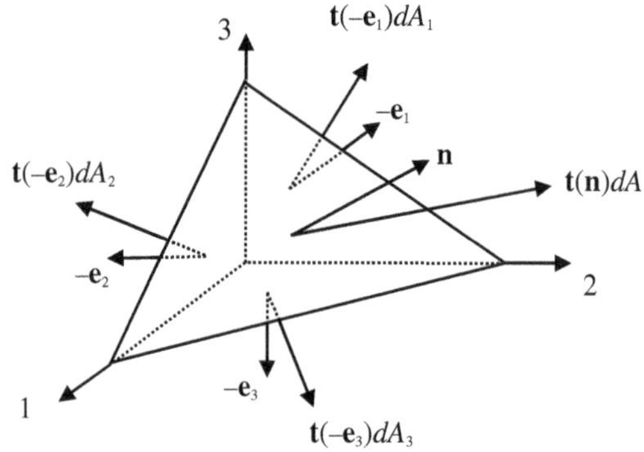

Fig. 3.3. Outward unit normal vectors and tractions on the surface of a tetrahedron.

or

$$\mathbf{n}dA = \mathbf{e}_j dA_j. \tag{3.22}$$

Dotting both sides of Eq. (3.22) by $\mathbf{e}_i$, we have

$$\mathbf{e}_i \cdot \mathbf{n}dA = \mathbf{e}_i \cdot \mathbf{e}_j dA_j, \tag{3.23}$$

$$n_i dA = \delta_{ij} dA_j, \tag{3.24}$$

$$n_i dA = dA_i. \tag{3.25}$$

We denote the tractions on the three faces of the tetrahedron with their outward normal directions along $-\mathbf{e}_i$ by

$$\begin{aligned} &\mathbf{t}(-\mathbf{e}_i) = -\mathbf{t}_i(\mathbf{e}_i) = -\mathbf{t}_i, \\ &\mathbf{t}_i = \mathbf{t}_i(\mathbf{e}_i), \end{aligned} \tag{3.26}$$

where Eq. (3.17) has been used. The application of the linear momentum equation in Eq. (3.2) to the tetrahedron gives

$$\mathbf{t}(\mathbf{n})dA + \mathbf{t}(-\mathbf{e}_1)dA_1 + \mathbf{t}(-\mathbf{e}_2)dA_2 + \mathbf{t}(-\mathbf{e}_3)dA_3 \cong 0, \tag{3.27}$$

where the body force and acceleration terms have been neglected as higher-order infinitesimals. With the use of Eq. (3.26), we can write

Eq. (3.27) as

$$\mathbf{t}(\mathbf{n})dA - \mathbf{t}(\mathbf{e}_1)dA_1 - \mathbf{t}(\mathbf{e}_2)dA_2 - \mathbf{t}(\mathbf{e}_3)dA_3 = 0, \tag{3.28}$$

$$\mathbf{t}(\mathbf{n})dA = \mathbf{t}_1 dA_1 + \mathbf{t}_2 dA_2 + \mathbf{t}_3 dA_3 = \mathbf{t}_j dA_j, \tag{3.29}$$

$$\mathbf{t}(\mathbf{n})dA = \mathbf{t}_j dA_j. \tag{3.30}$$

Using

$$dA_j = n_j dA, \tag{3.31}$$

we can write Eq. (3.30) as

$$\mathbf{t} = \mathbf{t}(\mathbf{n}) = \mathbf{t}_j n_j. \tag{3.32}$$

Next, we introduce the Cauchy stress tensor $\boldsymbol{\tau}$ by breaking $\mathbf{t}_j$ into components in terms of τ_{jk}:

$$\mathbf{t}_j = \tau_{jk}\mathbf{e}_k, \tag{3.33}$$

or

$$\begin{aligned}
\mathbf{t}_1 &= \tau_{11}\mathbf{e}_1 + \tau_{12}\mathbf{e}_2 + \tau_{13}\mathbf{e}_3, \\
\mathbf{t}_2 &= \tau_{21}\mathbf{e}_1 + \tau_{22}\mathbf{e}_2 + \tau_{23}\mathbf{e}_3, \\
\mathbf{t}_3 &= \tau_{31}\mathbf{e}_1 + \tau_{32}\mathbf{e}_2 + \tau_{33}\mathbf{e}_3.
\end{aligned} \tag{3.34}$$

The first index of τ_{kl} is associated with the normal of the plane on which $\mathbf{t}_k$ is acting. The second index of τ_{kl} is associated with the components of $\mathbf{t}_k$. τ_{kl} can be described by the following matrix:

$$[\tau_{kl}] = \begin{bmatrix} \tau_{11} & \tau_{12} & \tau_{13} \\ \tau_{21} & \tau_{22} & \tau_{23} \\ \tau_{31} & \tau_{32} & \tau_{33} \end{bmatrix}, \tag{3.35}$$

whose components are shown in Fig. 3.4.

With the stress tensor, we can write Eq. (3.32) using the component forms of $\mathbf{t}$ and $\mathbf{t}_i$ as

$$t_k\mathbf{e}_k = \tau_{jk}\mathbf{e}_k n_j. \tag{3.36}$$

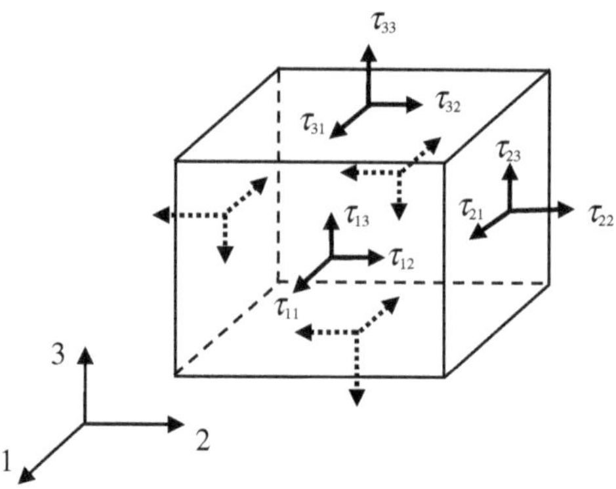

Fig. 3.4. Cauchy stress components.

Dotting both sides of Eq. (3.36) by $\mathbf{e}_i$, we obtain

$$t_i = n_j \tau_{ji}. \tag{3.37}$$

In matrix form, we have

$$\begin{bmatrix} t_1 & t_2 & t_3 \end{bmatrix} = \begin{bmatrix} n_1 & n_2 & n_3 \end{bmatrix} \begin{bmatrix} \tau_{11} & \tau_{12} & \tau_{13} \\ \tau_{21} & \tau_{22} & \tau_{23} \\ \tau_{31} & \tau_{32} & \tau_{33} \end{bmatrix}, \tag{3.38}$$

or

$$t_i = \tau_{ij}^T n_j, \tag{3.39}$$

and

$$\begin{bmatrix} t_1 \\ t_2 \\ t_3 \end{bmatrix} = \begin{bmatrix} \tau_{11} & \tau_{21} & \tau_{31} \\ \tau_{12} & \tau_{22} & \tau_{32} \\ \tau_{13} & \tau_{23} & \tau_{33} \end{bmatrix} \begin{bmatrix} n_1 \\ n_2 \\ n_3 \end{bmatrix}. \tag{3.40}$$

3.5 Linear Momentum Equation

From the linear momentum equation in Eq. (3.2), using Eq. (3.14) and

$$t_i = n_j \tau_{ji}, \tag{3.41}$$

we have

$$\int_v \rho \frac{dv_i}{dt} dv = \int_v \rho f_i dv + \int_s t_i ds$$
$$= \int_v \rho f_i dv + \int_s \tau_{ji} n_j ds = \int_v \rho f_i dv + \int_v \tau_{ji,j} dv. \qquad (3.42)$$

Hence,

$$\tau_{ji,j} + \rho f_i = \rho \dot{v}_i. \qquad (3.43)$$

3.6 Angular Momentum Equation

In terms of components, the angular momentum equation in Eq. (3.3) takes the following form:

$$\frac{d}{dt} \int_v \varepsilon_{ijk} y_j \rho v_k dv = \int_v \varepsilon_{ijk} y_j \rho f_k dv + \int_s \varepsilon_{ijk} y_j t_k ds. \qquad (3.44)$$

The term on the left-hand side of Eq. (3.44) can be written as

$$\frac{d}{dt} \int_v \varepsilon_{ijk} y_j \rho v_k dv = \int_v \rho \varepsilon_{ijk} \frac{d}{dt}(y_j v_k) dv$$
$$= \int_v \rho \varepsilon_{ijk} (\dot{y}_j v_k + y_j \dot{v}_k) dv$$
$$= \int_v \rho \varepsilon_{ijk} (v_j v_k + y_j \dot{v}_k) dv = \int_v \rho \varepsilon_{ijk} y_j \dot{v}_k dv. \qquad (3.45)$$

The last term on the right-hand side of Eq. (3.44) can be written as

$$\int_s \varepsilon_{ijk} y_j t_k ds = \int_s \varepsilon_{ijk} y_j \tau_{lk} n_l ds$$
$$= \int_v (\varepsilon_{ijk} y_j \tau_{lk})_{,l} dv = \int_v \varepsilon_{ijk} (\delta_{jl} \tau_{lk} + y_j \tau_{lk,l}) dv$$
$$= \int_v \varepsilon_{ijk} (\tau_{jk} + y_j \tau_{lk,l}) dv. \qquad (3.46)$$

Substituting Eqs. (3.45) and (3.46) into Eq. (3.44), we obtain

$$\int_v \rho\varepsilon_{ijk} y_j \dot{v}_k dv = \int_v \varepsilon_{ijk} y_j \rho f_k dv + \int_v \varepsilon_{ijk}(\tau_{jk} + y_j \tau_{lk,l}) dv, \quad (3.47)$$

or

$$\int_v \varepsilon_{ijk} y_j (\rho\dot{v}_k - \rho f_k - \tau_{lk,l}) dv = \int_v \varepsilon_{ijk}\tau_{jk} dv. \quad (3.48)$$

The left-hand side of Eq. (3.48) vanishes because of the linear momentum equation in Eq. (3.43). Hence,

$$\int_v \varepsilon_{ijk}\tau_{jk} dv = 0, \quad (3.49)$$

which implies that

$$\varepsilon_{ijk}\tau_{jk} = 0. \quad (3.50)$$

Equation (3.50) further implies that the Cauchy stress tensor τ_{kl} is symmetric, i.e., $\tau_{kl} = \tau_{lk}$. To see this, we multiply Eq. (3.50) by ε_{imn}. Then

$$\varepsilon_{imn}\varepsilon_{ijk}\tau_{jk} = (\delta_{jm}\delta_{kn} - \delta_{jn}\delta_{km})\,\tau_{jk} = \tau_{mn} - \tau_{nm} = 0. \quad (3.51)$$

3.7 Maximum Normal Stress

Consider an areal element at **y** with a normal **n** (see Fig. 3.5). The traction on the element is **t**. We have

$$t_i = n_j \tau_{ji}. \quad (3.52)$$

t has a normal component along **n** which is given by

$$p = \mathbf{t} \cdot \mathbf{n} = t_i n_i = n_j \tau_{ji} n_i. \quad (3.53)$$

Then the normal traction vector is

$$p_k = p n_k = t_i n_i n_k = n_j \tau_{ji} n_i n_k. \quad (3.54)$$

The shear stress vector on the areal element is given by

$$s_k = t_k - p_k = n_j \tau_{jk} - \tau_{ij} n_i n_j n_k. \quad (3.55)$$

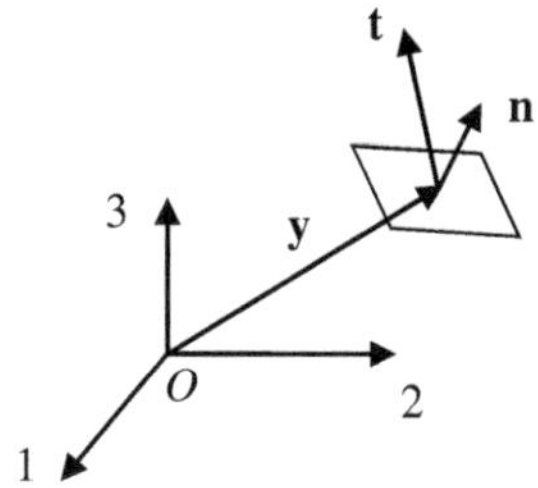

Fig. 3.5. Traction on an areal element at $\mathbf{y}$.

In this section, we maximize (or minimize)

$$p(\mathbf{n}) = n_j \tau_{ji} n_i \tag{3.56}$$

under the constraint

$$n_i n_i = 1. \tag{3.57}$$

For stationary values of p, we have

$$dp = (dn_j)\tau_{ji} n_i + n_j \tau_{ji}(dn_i) = 2n_j \tau_{ji}(dn_i) = 2t_i(dn_i) = 0, \tag{3.58}$$

or

$$\mathbf{t} \cdot (d\mathbf{n}) = 0. \tag{3.59}$$

Hence, $\mathbf{t}$ is perpendicular to any $d\mathbf{n}$. On the other hand, from Eq. (3.57),

$$d(n_i n_i) = 2n_i(dn_i) = 2\mathbf{n} \cdot (d\mathbf{n}) = 0, \tag{3.60}$$

or $\mathbf{n}$ is perpendicular to $d\mathbf{n}$. $d\mathbf{n}$ forms a plane perpendicular to $\mathbf{n}$. Hence $\mathbf{t}$ is parallel to $\mathbf{n}$ or $\mathbf{t} = \lambda\mathbf{n}$. In other words, $\mathbf{t}$ is along the normal, i.e., $\mathbf{t} = \mathbf{p}$. In this case, the shear stress $\mathbf{s} = 0$ and λ is the stationary p. Therefore,

$$t_i = n_j \tau_{ji} = \lambda n_i = \lambda \delta_{ij} n_j, \tag{3.61}$$

or

$$(\tau_{ij} - \lambda \delta_{ij}) n_j = 0. \tag{3.62}$$

Equation (3.62) is a system of linear and homogeneous equations for $\mathbf{n}$ with an undetermined parameter λ in the coefficient matrix (an eigenvalue problem). For nontrivial solutions of $\mathbf{n}$, we must have

$$\det(\tau_{ij} - \lambda\delta_{ij}) = 0, \tag{3.63}$$

which can be written as a cubic equation for λ:

$$\lambda^3 - J_1\lambda^2 + J_2\lambda - J_3 = 0. \tag{3.64}$$

J_1, J_2 and J_3 are three scalar invariants of $\boldsymbol{\tau}$:

$$\begin{aligned} J_1 &= \tau_{kk}, \\ J_2 &= \frac{1}{2}(\tau_{kk}\tau_{ll} - \tau_{kl}\tau_{lk}) \\ &= \begin{vmatrix} \tau_{11} & \tau_{12} \\ \tau_{21} & \tau_{22} \end{vmatrix} + \begin{vmatrix} \tau_{22} & \tau_{23} \\ \tau_{32} & \tau_{33} \end{vmatrix} + \begin{vmatrix} \tau_{33} & \tau_{31} \\ \tau_{13} & \tau_{11} \end{vmatrix}, \\ J_3 &= \det(\tau_{kl}). \end{aligned} \tag{3.65}$$

Once λ is known, we can solve Eq. (3.62) for $\mathbf{n}$. Since τ_{jk} is real and symmetric, there exist three real roots (eigenvalues) for λ which are called principal stresses and are denoted by

$$\tau^{(1)} \geq \tau^{(2)} \geq \tau^{(3)}. \tag{3.66}$$

The corresponding principal directions (eigenvectors) are denoted by

$$\mathbf{n}^{(1)}, \quad \mathbf{n}^{(2)}, \quad \mathbf{n}^{(3)}, \tag{3.67}$$

which are mutually orthogonal as shown in Fig. 3.6. Along the principle directions, the shear stress components of $\boldsymbol{\tau}$ vanish. In other words, $\boldsymbol{\tau}$ is diagonalized.

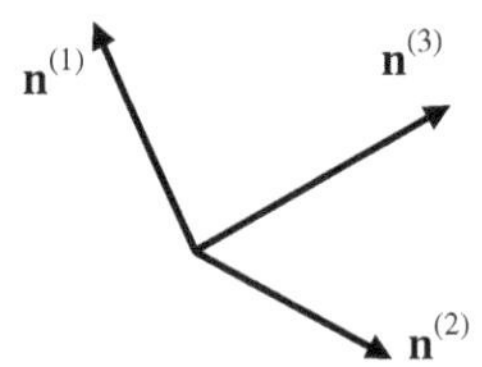

Fig. 3.6. Principal directions.

3.8 Maximum Shear Stress

We denote the principal directions of $\boldsymbol{\tau}$ by

$$\mathbf{e}_1 = \mathbf{n}^{(1)}, \quad \mathbf{e}_2 = \mathbf{n}^{(2)}, \quad \mathbf{e}_3 = \mathbf{n}^{(3)}. \tag{3.68}$$

With respect to the principal directions, we can write

$$\boldsymbol{\tau} = \tau^{(1)}\mathbf{e}_1\mathbf{e}_1 + \tau^{(2)}\mathbf{e}_2\mathbf{e}_2 + \tau^{(3)}\mathbf{e}_3\mathbf{e}_3, \tag{3.69}$$

$$\begin{aligned}\mathbf{t} = \mathbf{n}\cdot\boldsymbol{\tau} &= (n_1\mathbf{e}_1 + n_2\mathbf{e}_2 + n_3\mathbf{e}_3)\cdot(\tau^{(1)}\mathbf{e}_1\mathbf{e}_1 + \tau^{(2)}\mathbf{e}_2\mathbf{e}_2 + \tau^{(3)}\mathbf{e}_3\mathbf{e}_3)\\ &= n_1\tau^{(1)}\mathbf{e}_1 + n_2\tau^{(2)}\mathbf{e}_2 + n_3\tau^{(3)}\mathbf{e}_3,\end{aligned} \tag{3.70}$$

$$\mathbf{t}\cdot\mathbf{t} = n_1^2(\tau^{(1)})^2 + n_2^2(\tau^{(2)}) + n_3^2(\tau^{(3)})^2, \tag{3.71}$$

$$\begin{aligned}p = \mathbf{t}\cdot\mathbf{n} &= (n_1\tau^{(1)}\mathbf{e}_1 + n_2\tau^{(2)}\mathbf{e}_2 + n_3\tau^{(3)}\mathbf{e}_3)\cdot(n_1\mathbf{e}_1 + n_2\mathbf{e}_2 + n_3\mathbf{e}_3)\\ &= \tau^{(1)}n_1^2 + \tau^{(2)}n_2^2 + \tau^{(3)}n_3^2,\end{aligned} \tag{3.72}$$

$$\begin{aligned}s^2 &= \mathbf{t}\cdot\mathbf{t} - p^2\\ &= n_1^2(\tau^{(1)})^2 + n_2^2(\tau^{(2)})^2 + n_3^2(\tau^{(3)})^2 - (\tau^{(1)}n_1^2 + \tau^{(2)}n_2^2 + \tau^{(3)}n_3^2)^2\\ &= (\tau^{(1)} - \tau^{(2)})^2n_1^2n_2^2 + (\tau^{(2)} - \tau^{(3)})^2n_2^2n_3^2 + (\tau^{(3)} - \tau^{(1)})^2n_3^2n_1^2.\end{aligned} \tag{3.73}$$

The three components of $\mathbf{n}$ are not independent because $\mathbf{n}\cdot\mathbf{n} = 1$. We use

$$n_3^2 = 1 - n_1^2 - n_2^2 \tag{3.74}$$

to write s^2 as a function of the two independent variables, i.e., n_1 and n_2:

$$\begin{aligned}s^2 &= (\tau^{(1)} - \tau^{(2)})^2n_1^2n_2^2 + (\tau^{(2)} - \tau^{(3)})^2n_2^2(1 - n_1^2 - n_2^2)\\ &\quad + (\tau^{(3)} - \tau^{(1)})^2n_1^2(1 - n_1^2 - n_2^2) = s^2(n_1, n_2).\end{aligned} \tag{3.75}$$

For stationary values of s^2, we set

$$\begin{aligned}\frac{\partial s^2}{\partial n_1} &= (\tau^{(1)} - \tau^{(2)})^2 2n_1n_2^2 + (\tau^{(2)} - \tau^{(3)})^2n_2^2(-2n_1)\\ &\quad + (\tau^{(3)} - \tau^{(1)})^2 2n_1(1 - n_1^2 - n_2^2) + (\tau^{(3)} - \tau^{(1)})^2n_1^2(-2n_1) = 0,\end{aligned} \tag{3.76}$$

$$\frac{\partial s^2}{\partial n_2} = (\tau^{(1)} - \tau^{(2)})^2 n_1^2 2n_2 + (\tau^{(3)} - \tau^{(1)})^2 n_1^2(-2n_2)$$
$$+ (\tau^{(2)} - \tau^{(3)})^2 2n_2(1 - n_1^2 - n_2^2)$$
$$+ (\tau^{(2)} - \tau^{(3)})^2 n_2^2(-2n_2) = 0. \tag{3.77}$$

n_1 is a common factor of Eq. (3.76). n_2 is a common factor of Eq. (3.77). There are four possibilities. In case (i), $n_1 = n_2 = 0$. In this case, $n_3 = \pm 1$, which is a principal direction. We have $s^2 = 0$, which is a minimum. In case (ii), $n_1 = 0$ and $n_2 \neq 0$. Equation (3.76) becomes trivial, and Eq. (3.77) reduces to

$$(\tau^{(2)} - \tau^{(3)})^2 2n_2(1 - n_2^2) + (\tau^{(2)} - \tau^{(3)})^2 n_2^2(-2n_2) = 0. \tag{3.78}$$

Assuming $\tau^{(2)} \neq \tau^{(3)}$, we obtain from Eq. (3.78) that

$$n_2 = \pm 1/\sqrt{2}. \tag{3.79}$$

Then, from Eq. (3.74),

$$n_3 = \pm 1/\sqrt{2}, \tag{3.80}$$

$$\mathbf{n} = (0, \pm 1/\sqrt{2}, \pm 1/\sqrt{2}), \tag{3.81}$$

$$s^2 = \frac{1}{4}(\tau^{(2)} - \tau^{(3)})^2, \quad s = \frac{\tau^{(2)} - \tau^{(3)}}{2}. \tag{3.82}$$

Similarly, in case (iii) with $n_1 \neq 0$ and $n_2 = 0$, Eq. (3.77) becomes trivial and Eq. (3.76) reduces to

$$(\tau^{(3)} - \tau^{(1)})^2 2n_1(1 - n_1^2) + (\tau^{(3)} - \tau^{(1)})^2 n_1^2(-2n_1) = 0. \tag{3.83}$$

Assuming $\tau^{(1)} \neq \tau^{(3)}$, we have

$$n_1 = \pm 1/\sqrt{2}, \tag{3.84}$$

$$n_3 = \pm 1/\sqrt{2}, \tag{3.85}$$

$$\mathbf{n} = (\pm 1/\sqrt{2}, 0, \pm 1/\sqrt{2}), \tag{3.86}$$

$$s^2 = \frac{1}{4}(\tau^{(3)} - \tau^{(1)})^2, \quad s = \frac{\tau^{(1)} - \tau^{(3)}}{2}. \tag{3.87}$$

In case (iv) with $n_1 \neq 0$ and $n_2 \neq 0$, Eqs. (3.76) and (3.77) have no solution. If the following expression is used instead of Eq. (3.74):

$$n_2^2 = 1 - n_1^2 - n_3^2, \tag{3.88}$$

we have

$$s^2 = s^2(n_1, n_3), \tag{3.89}$$

from which another stationary value of s^2 can be found as

$$\mathbf{n} = (\pm 1/\sqrt{2}, \pm 1/\sqrt{2}, 0), \tag{3.90}$$

$$s^2 = \frac{1}{4}(\tau^{(2)} - \tau^{(1)})^2, \quad s = \frac{\tau^{(1)} - \tau^{(2)}}{2}. \tag{3.91}$$

3.9 Energy Equation

In terms of components, the conservation of energy in Eq. (3.4) takes the following form:

$$\begin{aligned}&\frac{d}{dt}\int_v \rho\left(\frac{1}{2}v_i v_i + \varepsilon\right)dv\\&\quad = \int_v (\rho f_k v_k + \rho r)dv + \int_s (t_k v_k - q_i n_i)ds.\end{aligned} \tag{3.92}$$

The left-hand side of Eq. (3.92) can be written as

$$\begin{aligned}&\frac{d}{dt}\int_v \rho\left(\frac{1}{2}v_i v_i + \varepsilon\right)dv = \int_v \rho\frac{d}{dt}\left(\frac{1}{2}v_i v_i + \varepsilon\right)dv\\&\quad = \int_v \rho\left(v_i \dot{v}_i + \dot{\varepsilon}\right)dv.\end{aligned} \tag{3.93}$$

The last term on the right-hand side of Eq. (3.92) can be written as

$$\begin{aligned}&\int_s (t_k v_k - q_i n_i)da = \int_s (\tau_{lk} n_l v_k - q_i n_i)ds\\&\quad = \int_v [(\tau_{lk} v_k)_{,l} - q_{i,i}]dv = \int_v [\tau_{lk,l} v_k + \tau_{lk} v_{k,l} - q_{i,i}]dv.\end{aligned} \tag{3.94}$$

The substitution of Eqs. (3.93) and (3.94) into Eq. (3.92) gives

$$\int_v \rho\left(v_k \dot{v}_k + \dot{\varepsilon}\right) dv$$
$$= \int_v (\rho f_k v_k + \rho r) dv + \int_v (\tau_{lk,l} v_k + \tau_{lk} v_{k,l} - q_{i,i}) dv, \tag{3.95}$$

or

$$\int_v v_k \left(\rho \dot{v}_k - \rho f_k - \tau_{lk,l}\right) dv = \int_v (\tau_{lk} v_{k,l} + \rho r - q_{i,i} - \rho\dot{\varepsilon}) dv. \tag{3.96}$$

The left-hand side of Eq. (3.96) vanishes because of the linear momentum equation in Eq. (3.43). Hence,

$$\int_v (\tau_{lk} v_{k,l} + \rho r - q_{i,i} - \rho\dot{\varepsilon}) dv = 0, \tag{3.97}$$

which implies that

$$\rho\dot{\varepsilon} = \tau_{ij} v_{j,i} + \rho r - q_{i,i}. \tag{3.98}$$

3.10 Entropy Inequality

Using the divergence theorem and Eq. (3.13), we can convert the entropy inequality in Eq. (3.5) into its differential form as

$$\rho\dot{\eta} \geq \frac{\rho r}{\theta} - \left(\frac{q_i}{\theta}\right)_{,i}. \tag{3.99}$$

It can be further written as

$$\rho r \leq \theta \left(\frac{q_i}{\theta}\right)_{,i} + \theta\rho\frac{d\eta}{dt} = \theta\left(\frac{q_{i,i}}{\theta} - \frac{q_i}{\theta^2}\theta_{,i}\right) + \theta\rho\frac{d\eta}{dt}$$
$$= q_{i,i} - \frac{q_i}{\theta}\theta_{,i} + \theta\rho\frac{d\eta}{dt}. \tag{3.100}$$

Eliminating r from Eqs. (3.98) and (3.100), we obtain the C–D (Clausius–Duhem) inequality as

$$\rho(\theta\dot{\eta} - \dot{\varepsilon}) + \tau_{ij} v_{j,i} - \frac{q_i \theta_{,i}}{\theta} \geq 0. \tag{3.101}$$

A free energy density χ per unit mass can be introduced through the following Legendre transform:

$$\chi = \varepsilon - \theta\eta. \tag{3.102}$$

Then the energy equation in Eq. (3.98) and the C–D inequality in Eq. (3.101) become

$$\rho(\dot{\chi} + \eta\dot{\theta} + \dot{\eta}\theta) = \tau_{ij}v_{j,i} + \rho r - q_{i,i}, \tag{3.103}$$

and

$$-\rho(\dot{\chi} + \eta\dot{\theta}) + \tau_{ij}v_{j,i} - \frac{q_i\theta_{,i}}{\theta} \geq 0. \tag{3.104}$$

3.11 Summary of Balance Laws

For convenience, we gather the differential forms of the balance laws from Eqs. (3.10), (3.43), (3.50), (3.98) and (3.99) as follows:

$$\dot{\rho} + \rho v_{i,i} = 0, \tag{3.105}$$

$$\rho\dot{v}_k = \tau_{ik,i} + \rho f_k, \tag{3.106}$$

$$\varepsilon_{ijk}\tau_{jk} = 0, \tag{3.107}$$

$$\rho\dot{\varepsilon} = \tau_{mk}v_{k,m} + \rho r - q_{i,i}, \tag{3.108}$$

$$\rho\dot{\eta} \geq \frac{\rho r}{\theta} - \left(\frac{q_i}{\theta}\right)_{,i}. \tag{3.109}$$

Equations (3.105)–(3.109) were derived from their integral forms in Eqs. (3.1)–(3.5) which are summarize into the following unified form:

$$\frac{d}{dt}\int_v \rho\Psi dv = \int_v \rho\Gamma dv + \int_s \mathbf{n}\cdot\mathbf{\Phi} ds, \tag{3.110}$$

where, for the entropy inequality, "=" should be replaced by "≥". The conservations of mass, linear momentum, angular momentum, energy and the entropy inequality correspond to

	Ψ	Γ	$\mathbf{\Phi}$
Mass	1	0	0
Linear momentum	$\mathbf{v}$	$\mathbf{f}$	$\boldsymbol{\tau}$
Angular momentum	$\mathbf{y} \times \mathbf{v}$	$\mathbf{y} \times \mathbf{f}$	$-\boldsymbol{\tau} \times \mathbf{y}$
Energy	$\varepsilon + \mathbf{v} \cdot \mathbf{v}/2$	$\mathbf{f} \cdot \mathbf{v} + r$	$\boldsymbol{\tau} \cdot \mathbf{v} - \mathbf{q}$
Entropy	η	r/θ	$-\mathbf{q}/\theta$

(3.111)

When the volume v does not have a discontinuity surface, using the conventional transport theorem in Eq. (2.83) (or the material time derivative in Eq. (3.13)) and the conventional divergence theorem in Eq. (1.136), we obtain the differential form of Eq. (3.110) as

$$\rho \frac{d\Psi}{dt} = \rho\Gamma + \nabla \cdot \mathbf{\Phi}, \tag{3.112}$$

where Eq. (3.13) has been used.

3.12 Jump Conditions

The field equations in Eqs. (3.105)–(3.109) in differential forms were obtained by applying the integral forms of the balance laws in Eqs. (3.1)–(3.5) to a material body in which the fields were continuous and differentiable. When there is a discontinuity surface in the material body, the implications of the integral balance laws need to be re-examined. When a discontinuity surface Σ is present in v as shown in Fig. 3.7, we need to use the modified transport theorem in Eq. (2.104) and the modified divergence theorem in Eq. (1.149).

From Eq. (2.104), with $\phi = \rho\Psi$, we have

$$\frac{d}{dt}\int_v \rho\Psi dv = \int_{v-\Sigma} \left\{ \frac{\partial(\rho\Psi)}{\partial t} + (\rho\Psi v_k)_{,k} \right\} dv - \int_\Sigma [\rho\Psi\Theta] ds. \tag{3.113}$$

From the modified divergence theorem in Eq. (1.149),

$$\int_s \mathbf{n} \cdot \mathbf{\Phi} ds = \int_{v-\Sigma} \nabla \cdot \mathbf{\Phi} dv + \int_\Sigma \mathbf{n} \cdot [\mathbf{\Phi}] ds. \tag{3.114}$$

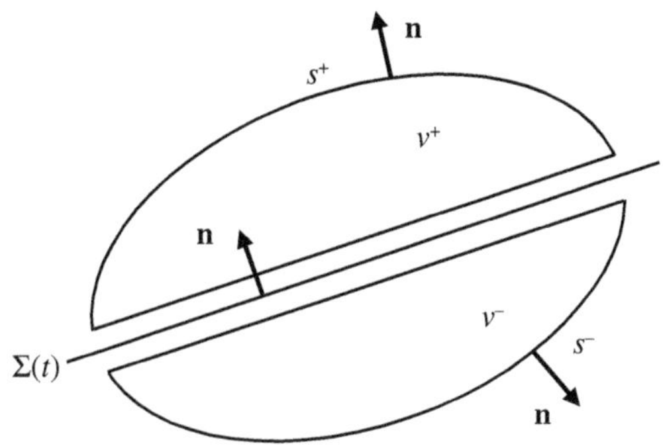

Fig. 3.7. A material body with a discontinuity surface.

The substitutions of Eqs. (3.113) and (3.114) into the general balance law in Eq. (3.110) gives

$$\int_{v-\Sigma}\left\{\frac{\partial(\rho\Psi)}{\partial t}+(\rho\Psi v_k)_{,k}\right\}dv-\int_{\Sigma}[\rho\Psi\Theta]ds$$
$$=\int_{v}\rho\Gamma dv+\int_{v-\Sigma}\nabla\cdot\mathbf{\Phi}dv+\int_{\Sigma}\mathbf{n}\cdot[\mathbf{\Phi}]ds, \qquad (3.115)$$

or

$$\int_{v-\Sigma}\left\{\frac{\partial(\rho\Psi)}{\partial t}+(\rho\Psi v_k)_{,k}-\rho\Gamma-\nabla\cdot\mathbf{\Phi}\right\}dv$$
$$=\int_{\Sigma}[\mathbf{n}\cdot\mathbf{\Phi}+\rho\Psi\Theta]ds, \qquad (3.116)$$

where $\rho\Gamma$ is assumed to be bounded in v so that its integration over v is the same as its integration over v–Σ. With the conservation of mass in Eq. (3.105) valid in v_1 and v_2, the two parts of v in Fig. 3.7, we can write Eq. (3.116) as

$$\int_{v-\Sigma}\left\{\rho\dot{\Psi}-\rho\Gamma-\nabla\cdot\mathbf{\Phi}\right\}dv=\int_{\Sigma}[\mathbf{n}\cdot\mathbf{\Phi}+\rho\Psi\Theta]ds. \qquad (3.117)$$

Equation (3.117) leads to Eq. (3.112) in $v-\Sigma$ or each of v_1 and v_2 as well as

$$\int_{\Sigma}[\rho\Psi\Theta+\mathbf{n}\cdot\mathbf{\Phi}]ds=0, \qquad (3.118)$$

where $\Theta = N - \mathbf{v} \cdot \mathbf{n}$ (see Eq. (2.88)). Equation (3.118) implies the following jump condition across Σ:

$$[\rho\Psi\Theta + \mathbf{n} \cdot \boldsymbol{\Phi}] = 0. \tag{3.119}$$

From Eqs. (3.119) and (3.111) we have, for the conservation of mass, linear momentum, angular momentum, energy and the entropy inequality, the following jump conditions across Σ:

$$[\rho\Theta] = 0, \tag{3.120}$$

$$[\rho\mathbf{v}\Theta + \mathbf{n} \cdot \boldsymbol{\tau}] = 0, \tag{3.121}$$

$$[\rho\mathbf{y} \times \mathbf{v}\Theta + \mathbf{n} \cdot (-\boldsymbol{\tau} \times \mathbf{y})] = 0, \tag{3.122}$$

$$[\rho(\varepsilon + \mathbf{v} \cdot \mathbf{v}/2)\Theta + \mathbf{n} \cdot (\boldsymbol{\tau} \cdot \mathbf{v} - \mathbf{q})] = 0, \tag{3.123}$$

$$[\mathbf{n} \cdot \mathbf{q}/\theta - \rho\eta\Theta] \geq 0. \tag{3.124}$$

We assume the continuity of $\mathbf{y}$ across the discontinuity surface. Then Eq. (3.122) is implied by Eq. (3.121).

In the special case when the discontinuity surface is material, $\Theta = 0$, and Eq. (3.120) becomes trivial. In this case, Eqs. (3.121) and (3.123) reduce to

$$[\mathbf{n} \cdot \boldsymbol{\tau}] = 0, \tag{3.125}$$

$$[\mathbf{n} \cdot \mathbf{q}] = 0. \tag{3.126}$$

Assuming θ is continuous across the material surface, Eq. (3.124) reduces to

$$[\mathbf{n} \cdot \mathbf{q}] \geq 0 \quad \text{on} \quad \Sigma, \tag{3.127}$$

which is ensured by Eq. (3.126).

3.13 Material Forms of Balance Laws

For convenience, we recapitulate the differential forms of the balance laws from Eqs. (3.105)–(3.109) are as follows:

$$\dot{\rho} + \rho v_{i,i} = 0, \tag{3.128}$$

$$\rho\dot{v}_k = \tau_{ik,i} + \rho f_k, \tag{3.129}$$

$$\varepsilon_{ijk}\tau_{jk} = 0, \tag{3.130}$$

$$\rho\dot{\varepsilon} = \tau_{mk}v_{k,m} + \rho r - q_{i,i}, \tag{3.131}$$

$$\rho\dot{\eta} \geq \frac{\rho r}{\theta} - \left(\frac{q_i}{\theta}\right)_{,i}. \tag{3.132}$$

The above equations are written in terms of the present coordinates y_i in the sense that all spatial derivatives are taken with respect to $\mathbf{y}$. In the case of a solid continuum, usually the reference state of the continuum is known, while the present state is not. Therefore, it is essential to have the equations written in terms of the reference coordinates X_K. For this purpose, we introduce the first Piola–Kirchhoff stress tensor $K_{Lj,}$ which is a two-point tensor, the second Piola–Kirchhoff stress tensor P_{KL}, the material temperature gradient Θ_K, and the material heat flux vector Q_K through

$$K_{Lj} = JX_{L,i}\tau_{ij}, \quad \tau_{ij} = J^{-1}y_{i,L}K_{Lj}, \tag{3.133}$$

$$P_{KL} = JX_{K,i}X_{L,j}\,\tau_{ij} = P_{LK}, \quad \tau_{ij} = J^{-1}y_{i,K}y_{j,L}P_{KL}, \tag{3.134}$$

$$\Theta_K = \theta_{,i}y_{i,K} = \theta_{,K}, \quad Q_K = JX_{K,i}q_i, \tag{3.135}$$

With K_{Lj}, we have

$$\begin{aligned}\int_s t_j ds &= \int_s n_i\tau_{ij}ds = \int_s \tau_{ij}ds_i \\ &= \int_S \tau_{ij}JX_{L,i}dS_L = \int_S \tau_{ij}JX_{L,i}N_L dS \\ &= \int_V (\tau_{ij}JX_{L,i})_{,L}dV = \int_V K_{Lj,L}dV.\end{aligned} \tag{3.136}$$

Then the linear momentum equation in Eq. (3.2) can be written as

$$\int_V \rho\dot{v}_j JdV = \int_V \rho f_j JdV + \int_V K_{Lj,L}dV, \tag{3.137}$$

which implies that

$$K_{Lj,L} + \rho^0 f_j = \rho^0\dot{v}_j. \tag{3.138}$$

The angular momentum equation does not have a spatial derivative. In fact, it will be automatically satisfied later when the τ_{kl} given by the constitutive relation from a rotationally invariant free energy density is a symmetric tensor (Noether's theorem).

For the conservation of energy in Eq. (3.131), we have

$$\begin{aligned}
\rho\dot{\varepsilon} &= \tau_{ij}v_{j,i} + \rho r - q_{i,i} \\
&= \tau_{ij}(d_{ij} + \Omega_{ij}) + \rho r - q_{i,i} = \tau_{ij}d_{ij} + \rho r - q_{i,i} \\
&= J^{-1}y_{i,K}y_{j,L}P_{KL}d_{ij} + \rho^0 J^{-1}r - (J^{-1}y_{i,K}Q_K)_{,i} \\
&= J^{-1}P_{KL}\dot{E}_{KL} + \rho^0 J^{-1}r - J^{-1}Q_{K,K},
\end{aligned} \tag{3.139}$$

where Eq. (2.77) has been used. Then

$$\rho^0\dot{\varepsilon} = P_{KL}\dot{E}_{KL} + \rho^0 r - Q_{K,K}. \tag{3.140}$$

In terms of the free energy

$$\chi = \varepsilon - \theta\eta, \tag{3.141}$$

the energy equation becomes

$$\rho^0(\dot{\chi} + \eta\dot{\theta} + \dot{\eta}\theta) = P_{KL}\dot{E}_{KL} + \rho^0 r - Q_{K,K}. \tag{3.142}$$

The C–D inequality in Eq. (3.104) takes the following form:

$$-\rho^0(\dot{\chi} + \eta\dot{\theta}) + P_{KL}\dot{E}_{KL} - \frac{Q_K\Theta_K}{\theta} \geq 0. \tag{3.143}$$

In summary, the balance laws in material forms are

$$\rho^0 = \rho J, \tag{3.144}$$

$$K_{Lk,L} + \rho^0 f_k = \rho^0\dot{v}_k, \tag{3.145}$$

$$\varepsilon_{kij}\tau_{ij} = 0, \tag{3.146}$$

$$\rho^0(\dot{\chi} + \eta\dot{\theta} + \dot{\eta}\theta) = P_{KL}\dot{E}_{KL} + \rho^0 r - Q_{K,K}, \tag{3.147}$$

$$-\rho^0(\dot{\chi} + \eta\dot{\theta}) + P_{KL}\dot{E}_{KL} - \frac{Q_K\Theta_K}{\theta} \geq 0, \tag{3.148}$$

where the spatial derivatives are taken with respect to $\mathbf{X}$.

Problems

3.1. Let v be a fixed spatial region. Show that Eqs. (3.1), (3.2) and (3.3) are equivalent to

$$\frac{\partial}{\partial t}\int_v \rho dv + \int_v (\mathbf{n}\cdot\mathbf{v})\rho dv = 0,$$

$$\frac{\partial}{\partial t}\int_v \rho\mathbf{v} dv + \int_v (\mathbf{n}\cdot\mathbf{v})\rho\mathbf{v} dv = \int_v \rho\mathbf{f} dv + \int_s \mathbf{t} ds,$$

$$\frac{\partial}{\partial t}\int_v \mathbf{y}\times\rho\mathbf{v} dv + \int_v (\mathbf{n}\cdot\mathbf{v})\mathbf{y}\times\rho\mathbf{v} dv = \int_v \mathbf{y}\times\rho\mathbf{f} dv + \int_s \mathbf{y}\times\mathbf{t} ds.$$

3.2. Derive the differential form of the conservation of mass in Eq. $(3.10)_2$ by studying the mass flowing into and out of a fixed differential element in space.

3.3. Let an external agency add mass to a continuum at a rate of $\mu(\mathbf{y},t)$ per unit volume per unit time. Derive the corresponding equation for conservation of mass.

3.4. Derive the differential form of the linear momentum equation in Eq. (3.43) by studying a material differential element.

3.5. Show that the linear momentum equation can be written as

$$\frac{\partial}{\partial t}(\rho v_i) = \rho f_i + (\tau_{ji} - \rho v_j v_i)_{,j}.$$

3.6. Show the symmetry of the Cauchy stress tensor by studying a material differential element.

3.7. Derive the differential form of the angular momentum equation when distributed body moments m_i per unit volume act throughout a continuum.

3.8. From $K_{Lj} = JX_{L,i}\tau_{ij}$, show that $\tau_{kl} = J^{-1}y_{k,M}K_{Ml}$.

3.9. Express the linear momentum equation in terms of the second Piola–Kirchhoff stress tensor P_{KL}.

3.10. From the material form of the linear momentum equation in Eq. (3.138), obtain its spatial form in Eq. (3.129) using Eq. $(3.133)_1$.

3.11. The stress components at a point are given by

$$\begin{pmatrix} 500 & 500 & 800 \\ 500 & 1000 & -750 \\ 800 & -750 & -300 \end{pmatrix}.$$

The unit normal of a plane is $(1/2,\ 1/2,\ 1/\sqrt{2})$. Determine:

(a) the traction vector on the plane;
(b) the components of the normal part of the traction vector;
(c) the components of the tangential part of the traction vector.

3.12. Determine the principal directions and principal stresses given that

$$[\sigma_{ij}] = \begin{bmatrix} 0 & \tau & 0 \\ \tau & 0 & 0 \\ 0 & 0 & 0 \end{bmatrix}.$$

3.13. Derive Eq. (3.62) by maximizing the p in Eq. (3.56) with the constraint in Eq. (3.57) using a Lagrange multiplier.

3.14. From Eq. (3.63), show Eqs. (3.64) and (3.65).

3.15. Let $\mathbf{n}$ be a unit vector which is equally inclined to the principal directions of $\boldsymbol{\tau}$. The plane normal to $\mathbf{n}$ is known as the octahedral plane. Calculate the normal stress and the resultant shear stress on the octahedral plane in terms of the principal stress components.

Chapter 4

Constitutive Relations

The balance laws in the previous chapter are insufficient to form a well-defined initial-boundary-value problem mathematically. Constitutive relations describing the behaviors of the specific material under consideration are necessary. In this chapter, we develop constitutive relations of elastic solids and viscous fluids, including thermal couplings. With the specifications of various constitutive relations, the energy equation and the C–D inequality assume new forms accordingly.

4.1 Nonlinear Elastic Solids

From the energy equation in Eq. (3.108), we have

$$\rho\dot{\varepsilon} = \tau_{mk}v_{k,m} + \rho r - q_{i,i}. \tag{4.1}$$

Consider an adiabatic process with $\mathbf{q} = 0$ and $r = 0$. Equation (4.1) reduces to

$$\rho\dot{\varepsilon} = \tau_{ij}v_{j,i}. \tag{4.2}$$

With

$$v_{j,i} = X_{M,i}\frac{d}{dt}(y_{j,M}), \tag{4.3}$$

we write Eq. (4.2) as

$$\rho\frac{d\varepsilon}{dt} = \tau_{ij}X_{M,i}\frac{d}{dt}(y_{j,M}). \tag{4.4}$$

Motivated by Eq. (4.4), we let

$$\varepsilon = \varepsilon(y_{j,M}). \tag{4.5}$$

Then

$$\frac{d\varepsilon}{dt} = \frac{\partial \varepsilon}{\partial(y_{j,M})}\frac{d}{dt}(y_{j,M}). \tag{4.6}$$

Substituting Eq. (4.6) into Eq. (4.4), we obtain

$$\rho\frac{\partial \varepsilon}{\partial(y_{j,M})}\frac{d}{dt}(y_{j,M}) = \tau_{ij}X_{M,i}\frac{d}{dt}(y_{j,M}), \tag{4.7}$$

or

$$\left[X_{M,i}\tau_{ij} - \rho\frac{\partial \varepsilon}{\partial(y_{j,M})}\right]\frac{d}{dt}(y_{j,M}) = 0. \tag{4.8}$$

Equation (4.8) implies the following constitutive relation:

$$\tau_{ij} = \rho y_{i,M}\frac{\partial \varepsilon}{\partial(y_{j,M})}. \tag{4.9}$$

Constitutive relations are assumed to be form-invariant under the change of observer in Eq. (2.106) (principle of objectivity [5]) for which, according to a theorem due to Cauchy (see Section 1.9 in Chapter 1), ε can only be a function of the following inner products of $y_{i,K}$ which are three vectors with respect to the spatial coordinates $\mathbf{y}$:

$$C_{KL} = y_{i,K}y_{i,L} \tag{4.10}$$

and the Jacobian $J>0$ which is not independent to $\mathbf{C}$ (see Eq. (2.35)). Instead of $\mathbf{C}$, we will use the strain tensor E_{KL}:

$$E_{KL} = (C_{KL} - \delta_{KL})/2. \tag{4.11}$$

Therefore, we take

$$\varepsilon = \varepsilon(E_{KL}). \tag{4.12}$$

Then the constitutive relation in Eq. (4.9) becomes

$$\tau_{ij} = \rho y_{i,M} \frac{\partial \varepsilon}{\partial E_{ML}} y_{j,L} = \tau_{ji} \tag{4.13}$$

Alternatively, Eq. (4.2) may be written in its material form (see Eq. (3.140)) as

$$\rho^0 \frac{d\varepsilon}{dt} = P_{KL} \frac{dE_{KL}}{dt}. \tag{4.14}$$

Equation (4.14) suggests the following internal energy density directly:

$$\varepsilon = \varepsilon(E_{KL}). \tag{4.15}$$

The substitution of Eq. (4.15) into Eq. (4.14) gives

$$\left(P_{KL} - \rho^0 \frac{\partial \varepsilon}{\partial E_{KL}}\right) \dot{E}_{KL} = 0, \tag{4.16}$$

or

$$\begin{aligned}
&\left(P_{11} - \rho^0 \frac{\partial \varepsilon}{\partial E_{11}}\right) \dot{E}_{11} + \left(P_{22} - \rho^0 \frac{\partial \varepsilon}{\partial E_{22}}\right) \dot{E}_{22} \\
&\quad + \left(P_{33} - \rho^0 \frac{\partial \varepsilon}{\partial E_{33}}\right) \dot{E}_{33} \\
&\quad + \left(P_{12} - \rho^0 \frac{\partial \varepsilon}{\partial E_{12}}\right) \dot{E}_{12} + \left(P_{21} - \rho^0 \frac{\partial \varepsilon}{\partial E_{21}}\right) \dot{E}_{21} \\
&\quad + \left(P_{23} - \rho^0 \frac{\partial \varepsilon}{\partial E_{23}}\right) \dot{E}_{23} + \left(P_{32} - \rho^0 \frac{\partial \varepsilon}{\partial E_{32}}\right) \dot{E}_{32} \\
&\quad + \left(P_{31} - \rho^0 \frac{\partial \varepsilon}{\partial E_{31}}\right) \dot{E}_{31} + \left(P_{13} - \rho^0 \frac{\partial \varepsilon}{\partial E_{13}}\right) \dot{E}_{13} = 0,
\end{aligned} \tag{4.17}$$

which can be further written as

$$\begin{aligned}
&\left(P_{11} - \rho^0 \frac{\partial \varepsilon}{\partial E_{11}}\right) \dot{E}_{11} + \left(P_{22} - \rho^0 \frac{\partial \varepsilon}{\partial E_{22}}\right) \dot{E}_{22} \\
&\quad + \left(P_{33} - \rho^0 \frac{\partial \varepsilon}{\partial E_{33}}\right) \dot{E}_{33} \\
&\quad + \left(P_{12} + P_{21} - \rho^0 \frac{\partial \varepsilon}{\partial E_{12}} - \rho^0 \frac{\partial \varepsilon}{\partial E_{21}}\right) \dot{E}_{12} \\
&\quad + \left(P_{23} + P_{32} - \rho^0 \frac{\partial \varepsilon}{\partial E_{23}} - \rho^0 \frac{\partial \varepsilon}{\partial E_{32}}\right) \dot{E}_{23} \\
&\quad + \left(P_{31} + P_{13} - \rho^0 \frac{\partial \varepsilon}{\partial E_{31}} - \rho^0 \frac{\partial \varepsilon}{\partial E_{13}}\right) \dot{E}_{31} = 0. \qquad (4.18)
\end{aligned}$$

Equation (4.18) is linear in $\dot{E}_{KL}$. For the equation to hold for any $\dot{E}_{KL} = \dot{E}_{LK}$, we must have

$$P_{KL} = \rho^0 \frac{1}{2}\left(\frac{\partial \varepsilon}{\partial E_{KL}} + \frac{\partial \varepsilon}{\partial E_{LK}}\right) = \rho^0 \frac{\partial \varepsilon}{\partial E_{KL}} = P_{LK}, \qquad (4.19)$$

where the partial derivatives with respect to E_{KL} are taken as if the strain components were independent, and ε is written as a symmetric function of E_{KL}, i.e.,

$$\begin{aligned}
&\frac{\partial E_{KL}}{\partial E_{LK}} = 0, \quad K \neq L, \\
&\frac{\partial \varepsilon}{\partial E_{KL}} = \frac{\partial \varepsilon}{\partial E_{LK}}.
\end{aligned} \qquad (4.20)$$

Then, from Eq. (3.133),

$$\begin{aligned}
K_{Lj} &= J X_{L,i} \tau_{ij} = J X_{L,i} (J^{-1} y_{i,K} y_{j,M} P_{KM}) \\
&= y_{j,M} P_{ML} = y_{j,K} \rho^0 \frac{\partial \varepsilon}{\partial E_{KL}}, \qquad (4.21)
\end{aligned}$$

$$\tau_{ij} = J^{-1} y_{i,K} y_{j,L} \rho^0 \frac{\partial \varepsilon}{\partial E_{KL}} = y_{i,K} y_{j,L} \rho \frac{\partial \varepsilon}{\partial E_{KL}} = \tau_{ji}. \qquad (4.22)$$

The internal energy density ε that determines the constitutive relations of nonlinear elastic materials may be written as

$$
\begin{aligned}
\rho^0\varepsilon(\mathbf{E}) &= \rho^0\varepsilon(0) + c_{1AB}E_{AB} \\
&\quad + \frac{1}{2}c_{2ABCD}E_{AB}E_{CD} + \frac{1}{6}c_{3ABCDEF}E_{AB}E_{CD}E_{EF} \\
&\quad + \frac{1}{24}c_{4ABCDEFGH}E_{AB}E_{CD}E_{EF}E_{GH} + \cdots ,
\end{aligned}
\tag{4.23}
$$

which produces

$$
\begin{aligned}
P_{KL} &= c_{1KL} + c_{2KLCD}E_{CD} + \frac{1}{2}c_{3KLCDEF}E_{CD}E_{EF} \\
&\quad + \frac{1}{6}c_{4KLCDEFGH}E_{CD}E_{EF}E_{GH} + \cdots .
\end{aligned}
\tag{4.24}
$$

The constant term in Eq. (4.23) is immaterial. The term linear in $\mathbf{E}$ in Eq. (4.23) describes initial stress which will be dropped below. The material constants in Eq. (4.23),

$$
c_{2ABCD}, \quad c_{3ABCDEF}, \quad c_{4ABCDEFGH}, \tag{4.25}
$$

are the second-, third- and fourth-order fundamental elastic constants, respectively. The second-order constants are mainly responsible for linear material behaviors. The third- and higher-order material constants are for nonlinear behaviors. For weak nonlinearity, the higher-order terms in Eq. (4.23) may be neglected. The structure of $\varepsilon(\mathbf{E})$ depends on material symmetry.

When a material is isotropic, the functional relationship of $\varepsilon(\mathbf{E})$ is invariant under the following orthogonal transformations (material isotropy):

$$
\begin{aligned}
\mathbf{X} &\to \mathbf{X}' = \mathbf{QX}, \\
\mathbf{E} &\to \mathbf{E}' = \mathbf{QEQ}^T,
\end{aligned}
\tag{4.26}
$$

i.e.,

$$
\varepsilon(\mathbf{E}) = \varepsilon(\mathbf{E}'), \tag{4.27}
$$

where

$$
\mathbf{E} = E_{KL}\mathbf{I}_K\mathbf{I}_L. \tag{4.28}
$$

In this case ε can be written as (see Prob. (1.18)):

$$\varepsilon = \varepsilon(I_1,\ I_2,\ I_3), \tag{4.29}$$

where

$$\begin{aligned} I_1 &= \mathrm{tr}(\mathbf{E}) = E_{KK}, \\ I_2 &= \mathrm{tr}(\mathbf{E}^2) = E_{KL}E_{LK}, \\ I_3 &= \mathrm{tr}(\mathbf{E}^3) = E_{KL}E_{LM}E_{MK}. \end{aligned} \tag{4.30}$$

Since

$$\begin{aligned} \frac{\partial I_1}{\partial E_{KL}} &= \delta_{KL}, \quad \frac{\partial I_1}{\partial \mathbf{E}} = \mathbf{1}, \\ \frac{\partial I_2}{\partial E_{KL}} &= 2E_{KL}, \quad \frac{\partial I_2}{\partial \mathbf{E}} = 2\mathbf{E}, \\ \frac{\partial I_3}{\partial E_{KL}} &= 3E_{LM}E_{MK}, \quad \frac{\partial I_3}{\partial \mathbf{E}} = 3\mathbf{E}^2, \end{aligned} \tag{4.31}$$

we have

$$P_{KL}\mathbf{I}_K\mathbf{I}_L = \rho^0\frac{\partial \varepsilon}{\partial I_1}\mathbf{1} + 2\rho^0\frac{\partial \varepsilon}{\partial I_2}\mathbf{E} + 3\rho^0\frac{\partial \varepsilon}{\partial I_3}\mathbf{E}^2. \tag{4.32}$$

4.2 Linear Elastic Solids

Consider small displacement gradients, i.e.,

$$||u_{i,K}|| << 1. \tag{4.33}$$

The deformation gradient tensor consists of zero- and first-order terms of the small displacement gradient and is kept as it is. However, when the deformation gradient is multiplied to an infinitesimal field, we may make the approximation that

$$F_{iK} = y_{i,K} = (\delta_{iL}X_L + u_i)_{,K} = \delta_{iK} + u_{i,K} \cong \delta_{iK}. \tag{4.34}$$

The Jacobian may be approximated by

$$
\begin{aligned}
J = \det(y_{i,K}) &= \begin{vmatrix} 1+u_{1,1} & u_{1,2} & u_{1,3} \\ u_{2,1} & 1+u_{2,2} & u_{2,3} \\ u_{3,1} & u_{3,2} & 1+u_{3,3} \end{vmatrix} \\
&\cong (1+u_{1,1})(1+u_{2,2})(1+u_{3,3}) \\
&\cong 1+u_{1,1}+u_{2,2}+u_{3,3} = 1+u_{K,K},
\end{aligned}
\tag{4.35}
$$

or, when multiplied to an infinitesimal field,

$$J \cong 1+u_{K,K} \cong 1. \tag{4.36}$$

The linear strain is denoted by $\boldsymbol{\varepsilon}$ according to Eq. (2.45):

$$E_{KL} \cong \frac{1}{2}(u_{L,K}+u_{K,L}) = \varepsilon_{KL}. \tag{4.37}$$

For small stresses, we have

$$K_{Lj} \cong \delta_{Li}\tau_{ij}, \quad P_{KL} \cong \delta_{Ki}\delta_{Lj}\tau_{ij}. \tag{4.38}$$

Since the stress tensors are approximately the same in the linear theory, we use σ_{ij} to denote the infinitesimal stress tensor in linear elasticity, i.e.,

$$
\begin{aligned}
K_{Lj} &\cong \delta_{Li}\tau_{ij} \to \sigma_{lj}, \\
P_{KL} &\cong \delta_{Ki}\delta_{Lj}\tau_{ij} \to \sigma_{kl},
\end{aligned}
\tag{4.39}
$$

where, as what is usually practiced in linear elasticity, only lowercase indices are used.

We also introduce the internal energy density U per unit volume as follows:

$$U = \rho^0\varepsilon \cong \frac{1}{2}{}^{c}{}_{2ABCD}\varepsilon_{AB}\varepsilon_{CD} \to \frac{1}{2}c_{ijkl}\varepsilon_{ij}\varepsilon_{kl}, \tag{4.40}$$

The linear constitutive relation generated by U is

$$\sigma_{ij} = \frac{\partial U}{\partial \varepsilon_{ij}} = c_{ijkl}\varepsilon_{kl}. \tag{4.41}$$

The elastic stiffness c_{ijkl} has the following symmetries:

$$c_{ijkl} = c_{jikl} = c_{klij}. \tag{4.42}$$

We also assume that the elastic stiffness is positive definite in the following sense:

$$\begin{aligned} &c_{ijkl}\varepsilon_{ij}\varepsilon_{kl} \geq 0 \quad \text{for any} \quad \varepsilon_{ij} = \varepsilon_{ji}, \\ &\text{and} \quad c_{ijkl}\varepsilon_{ij}\varepsilon_{kl} = 0 \quad \Rightarrow \quad \varepsilon_{ij} = 0. \end{aligned} \tag{4.43}$$

The linear constitutive relations in Eq. (4.41) can also be written as

$$\varepsilon_{ij} = s_{ijkl}\sigma_{kl}, \tag{4.44}$$

where s_{ijkl} is the elastic compliance.

We now introduce a compact matrix notation [12, 13]. It allows us to use matrices to represent the stress, strain, elastic stiffness and compliance tensors and is convenient for constitutive relations of anisotropic materials. This notation consists of replacing pairs of tensor indices ij or kl by single matrix indices p or q, where i, j, k and l take the values of 1, 2 and 3. p and q take the values of 1, 2, 3, 4, 5 and 6 according to

$$\begin{array}{lcccccc} ij \text{ or } kl: & 11 & 22 & 33 & 23 \text{ or } 32 & 31 \text{ or } 13 & 12 \text{ or } 21 \\ p \text{ or } q: & 1 & 2 & 3 & 4 & 5 & 6 \end{array} \tag{4.45}$$

Thus, for the stress tensor, we write

$$\begin{aligned} &\sigma_1 = \sigma_{11}, \quad \sigma_2 = \sigma_{22}, \quad \sigma_3 = \sigma_{33}, \\ &\sigma_4 = \sigma_{23}, \quad \sigma_5 = \sigma_{31}, \quad \sigma_6 = \sigma_{12}. \end{aligned} \tag{4.46}$$

For the strain tensor, we introduce ε_p such that

$$\begin{aligned} &\varepsilon_1 = \varepsilon_{11}, \quad \varepsilon_2 = \varepsilon_{22}, \quad \varepsilon_3 = \varepsilon_{33}, \\ &\varepsilon_4 = 2\varepsilon_{23}, \quad \varepsilon_5 = 2\varepsilon_{31}, \quad \varepsilon_6 = 2\varepsilon_{12}. \end{aligned} \tag{4.47}$$

Accordingly, the strain energy density per unit volume can be written as

$$U(\varepsilon) = \frac{1}{2}\sigma_p\varepsilon_p = \frac{1}{2}c_{pq}\varepsilon_p\varepsilon_q. \tag{4.48}$$

The corresponding constitutive relations take the following form:

$$\begin{aligned} \sigma_p &= \frac{\partial U}{\partial \varepsilon_p} = c_{pq}\varepsilon_q, \quad c_{pq} = c_{qp}, \\ \varepsilon_p &= s_{pq}\sigma_q, \quad s_{pq} = s_{qp}. \end{aligned} \tag{4.49}$$

The elastic properties of a fully anisotropic material such as a triclinic crystal are represented by the following full array of $[c_{pq}]$ with twenty one independent material constants:

$$\begin{bmatrix} \sigma_1 \\ \sigma_2 \\ \sigma_3 \\ \sigma_4 \\ \sigma_5 \\ \sigma_6 \end{bmatrix} = \begin{bmatrix} c_{11} & c_{12} & c_{13} & c_{14} & c_{15} & c_{16} \\ c_{21} & c_{22} & c_{23} & c_{24} & c_{25} & c_{26} \\ c_{31} & c_{32} & c_{33} & c_{34} & c_{35} & c_{36} \\ c_{41} & c_{42} & c_{43} & c_{44} & c_{45} & c_{46} \\ c_{51} & c_{52} & c_{53} & c_{54} & c_{55} & c_{56} \\ c_{61} & c_{62} & c_{63} & c_{64} & c_{65} & c_{66} \end{bmatrix} \begin{bmatrix} \varepsilon_1 \\ \varepsilon_2 \\ \varepsilon_3 \\ \varepsilon_4 \\ \varepsilon_5 \\ \varepsilon_6 \end{bmatrix}. \tag{4.50}$$

In the special case of an isotropic material with two independent material constants, the stress-strain relation reduces to

$$\begin{bmatrix} \sigma_1 \\ \sigma_2 \\ \sigma_3 \\ \sigma_4 \\ \sigma_5 \\ \sigma_6 \end{bmatrix} = \begin{bmatrix} c_{11} & c_{12} & c_{12} & 0 & 0 & 0 \\ c_{21} & c_{11} & c_{21} & 0 & 0 & 0 \\ c_{21} & c_{21} & c_{11} & 0 & 0 & 0 \\ 0 & 0 & 0 & c_{44} & 0 & 0 \\ 0 & 0 & 0 & 0 & c_{44} & 0 \\ 0 & 0 & 0 & 0 & 0 & c_{44} \end{bmatrix} \begin{bmatrix} \varepsilon_1 \\ \varepsilon_2 \\ \varepsilon_3 \\ \varepsilon_4 \\ \varepsilon_5 \\ \varepsilon_6 \end{bmatrix}, \tag{4.51}$$

or

$$\begin{aligned} \sigma_1 &= c_{11}\,\varepsilon_1 + c_{12}\,\varepsilon_2 + c_{12}\,\varepsilon_3, \\ \sigma_2 &= c_{21}\,\varepsilon_1 + c_{11}\,\varepsilon_2 + c_{12}\,\varepsilon_3, \\ \sigma_3 &= c_{21}\,\varepsilon_1 + c_{21}\,\varepsilon_2 + c_{11}\,\varepsilon_3, \\ \sigma_4 &= c_{44}\,\varepsilon_4, \quad \sigma_5 = c_{44}\,\varepsilon_5, \quad \sigma_6 = c_{44}\,\varepsilon_6, \end{aligned} \tag{4.52}$$

where

$$c_{44} = \frac{1}{2}(c_{11} - c_{12}). \tag{4.53}$$

For isotropic materials, using the symmetric fourth-order isotropic tensor in Eq. (1.168), we can write

$$c_{ijkl} = \lambda\delta_{ij}\delta_{kl} + \mu(\delta_{ik}\delta_{jl} + \delta_{il}\delta_{jk}), \tag{4.54}$$

where λ and μ are Lamé's elastic constants. Then,

$$\sigma_{ij} = c_{ijkl}\varepsilon_{kl} = \lambda\varepsilon_{kk}\delta_{ij} + 2\mu\varepsilon_{ij}. \tag{4.55}$$

The inverse of Eq. (4.55) is usually written as

$$\varepsilon_{ij} = \frac{1}{E}[(1+\nu)\sigma_{ij} - \nu\sigma_{kk}\delta_{ij}], \tag{4.56}$$

where E is Young's modulus and ν is Poisson's ratio. Contracting Eq. (4.55), we obtain

$$\sigma_{kk} = (3\lambda + 2\mu)\varepsilon_{kk}, \tag{4.57}$$

which can be written in terms of the mean normal stress as

$$\frac{1}{3}\sigma_{kk} = K\varepsilon_{kk}, \tag{4.58}$$

where

$$K = \lambda + \frac{2}{3}\mu. \tag{4.59}$$

K is the so-called bulk elastic modulus. From the positive definiteness of the strain energy density, it can be shown that $E > 0, -1 < \nu < 1/2$, $\mu > 0$ and $K > 0$. The relations of the constants c_{11} and c_{12} to Lamé's constants (λ, μ) and to Young's modulus, E, and Poisson's ratio, v, are given below [12]:

	c_{11}, c_{12}	λ, μ	E, ν
c_{11}	c_{11}	$\lambda + 2\mu$	$\frac{E(1-\nu)}{(1+\nu)(1-2\nu)}$
c_{12}	c_{12}	λ	$\frac{E\nu}{(1+\nu)(1-2\nu)}$
λ	c_{12}	λ	$\frac{E\nu}{(1+\nu)(1-2\nu)}$
μ	$\frac{c_{11}-c_{12}}{2}$	μ	$\frac{E}{2(1+\nu)}$
E	$\frac{(c_{11}+2c_{12})(c_{11}-c_{12})}{c_{11}+c_{12}}$	$\frac{\mu(3\lambda+2\mu)}{\lambda+\mu}$	E
ν	$\frac{c_{12}}{c_{11}+c_{12}}$	$\frac{\lambda}{2(\lambda+\mu)}$	ν

4.3 Nonlinear Thermoviscoelastic Solids

Thermal and dissipative effects often appear together and are treated in this section. We begin with the energy equation in Eq. (3.147) and the C–D inequality in Eq. (3.148) in material forms:

$$\rho^0(\dot{\chi} + \eta\dot{\theta} + \dot{\eta}\theta) = P_{KL}\dot{E}_{KL} + \rho^0 r - Q_{K,K}, \tag{4.60}$$

$$-\rho^0(\dot{\chi} + \eta\dot{\theta}) + P_{KL}\dot{E}_{KL} - \frac{Q_K\Theta_K}{\theta} \geq 0. \tag{4.61}$$

For constitutive relations we start with the following possible expressions:

$$\begin{aligned} \chi &= \chi(E_{KL};\ \theta;\ \Theta_K;\ \dot{E}_{KL}), \\ P_{KL} &= P_{KL}(E_{KL};\ \theta;\ \Theta_K;\ \dot{E}_{KL}), \\ Q_K &= Q_K(E_{KL};\ \theta;\ \Theta_K;\ \dot{E}_{KL}). \end{aligned} \tag{4.62}$$

The substitution of Eq. (4.62) into the C–D inequality in Eq. (4.61) yields

$$\begin{aligned} &-\rho^0\frac{\partial\chi}{\partial\Theta_K}\dot{\Theta}_K - \rho^0\left(\eta + \frac{\partial\chi}{\partial\theta}\right)\dot{\theta} - \rho^0\frac{\partial\chi}{\partial\dot{E}_{KL}}\ddot{E}_{KL} \\ &+\left(P_{KL} - \rho^0\frac{\partial\chi}{\partial E_{KL}}\right)\dot{E}_{KL} - \frac{1}{\theta}Q_K\Theta_K \geq 0. \end{aligned} \tag{4.63}$$

Since Eq. (4.63) is linear in $\dot{\Theta}_K$, $\dot{\theta}$ and $\ddot{E}_{KL}$, for the inequality to hold in general, we must have

$$\frac{\partial\chi}{\partial\Theta_K} = 0, \quad \eta + \frac{\partial\chi}{\partial\theta} = 0, \quad \frac{\partial\chi}{\partial\dot{E}_{KL}} = 0. \tag{4.64}$$

Hence,

$$\chi = \chi(E_{KL};\theta), \tag{4.65}$$

and Eq. (4.63) reduces to

$$\left(P_{KL} - \rho^0\frac{\partial\chi}{\partial E_{KL}}\right)\dot{E}_{KL} - \frac{1}{\theta}Q_K\Theta_K \geq 0. \tag{4.66}$$

The composition of Eq. (4.66) suggests the decomposition of **P** into recoverable and dissipative parts as

$$P_{KL} = P_{KL}^{R}(E_{KL};\theta) + P_{KL}^{D}(E_{KL};\theta;\Theta_K;\dot{E}_{KL}). \tag{4.67}$$

From Eq. (4.66), it then follows that

$$P_{KL}^{R} = \rho^0 \frac{\partial \chi}{\partial E_{KL}}, \tag{4.68}$$

$$P_{KL}^{D}\dot{E}_{KL} - \frac{1}{\theta}Q_K\Theta_K \geq 0. \tag{4.69}$$

Then, from Eq. (4.60), we obtain the heat or dissipation equation as

$$\rho^0\theta\dot{\eta} = P_{KL}^{D}\dot{E}_{KL} + \rho^0 r - Q_{K,K}. \tag{4.70}$$

The specific heat c under constant strain is defined by [14]

$$c = \left.\frac{\partial \varepsilon}{\partial \theta}\right|_{\mathbf{E}}. \tag{4.71}$$

With

$$\varepsilon = \chi + \eta\theta, \tag{4.72}$$

we have

$$c = \frac{\partial \chi}{\partial \theta} + \eta + \theta\frac{\partial \eta}{\partial \theta} = \theta\frac{\partial \eta}{\partial \theta}, \tag{4.73}$$

where Eq. (4.64) has been used.

4.4 Linear Thermoviscoelastic Solids

For small deformations, the recoverable part of the constitutive equations are determined by $\chi = \chi(\varepsilon_{ij}, \theta)$ and

$$\eta = -\frac{\partial \chi}{\partial \theta}, \quad \sigma_{ij}^{R} = \rho^0\frac{\partial \chi}{\partial \varepsilon_{ij}}. \tag{4.74}$$

In order to linearize the constitutive relations thermally, we expand χ into a power series about $\theta = T_0$ and ε_{ij} =0, where T_0 is a uniform reference temperature. Denoting $T = \theta - T_0$, assuming $|T/T_0| << 1$ and keeping quadratic terms only, we can write

$$\rho^0\chi = \frac{1}{2}c_{ijkl}\varepsilon_{ij}\varepsilon_{kl} - \frac{1}{2}\frac{\rho^0 c}{T_0}T^2 - \beta_{kl}\varepsilon_{kl}T, \tag{4.75}$$

where β_{ij} are the thermoelastic constants and c is the specific heat. Then Eqs. (4.74) produces

$$\begin{aligned} \sigma_{ij}^R &= c_{ijkl}\varepsilon_{kl} - \beta_{ij}T, \\ \rho^0\eta &= \frac{\rho^0 c}{T_0}T + \beta_{kl}\varepsilon_{kl}. \end{aligned} \tag{4.76}$$

Equations (4.73) and $(4.76)_2$ lead to

$$\theta\frac{\partial\eta}{\partial\theta} \cong T_0\frac{\partial\eta}{\partial T} = T_0\frac{1}{\rho^0}\frac{\rho^0 c}{T_0} = c. \tag{4.77}$$

For the dissipative part of the constitutive relations, we may choose, e.g., the following simple relations:

$$\sigma_{ij}^D = \mu_{ijkl}\dot{\varepsilon}_{kl}, \tag{4.78}$$

$$q_k = -\kappa_{kl}T_{,l}. \tag{4.79}$$

Equation (4.79) is Fourier's law for heat conduction. Equations (4.78) and (4.79) are restricted by

$$\sigma_{ij}^D\dot{\varepsilon}_{ij} - \frac{q_i T_{,i}}{T_0} \geq 0. \tag{4.80}$$

The linearized heat or dissipation equation is

$$\rho^0 T_0\dot{\eta} = \sigma_{ij}^D\dot{\varepsilon}_{ij} + \rho^0 r - q_{i,i}. \tag{4.81}$$

A dissipation function can be introduced as follows:

$$\Psi(\dot{\varepsilon}_{kl}) = \frac{1}{2}\mu_{ijkl}\dot{\varepsilon}_{kl}\dot{\varepsilon}_{ij}. \tag{4.82}$$

Then Eqs. (4.81), (4.78) and (4.80) can be written as

$$\rho^0 T_0\dot{\eta} = 2\Psi + \rho^0 r - q_{i,i}, \tag{4.83}$$

$$\sigma_{ij}^D = \frac{\partial \Psi}{\partial \dot{\varepsilon}_{ij}}, \tag{4.84}$$

$$2\Psi + \frac{\kappa_{ij} T_{,i} T_{,j}}{T_0} \geq 0. \tag{4.85}$$

To ensure that Eq. (4.85) is satisfied, we may require that

$$\Psi(\dot{\varepsilon}_{kl}) \geq 0, \quad \kappa_{ij}\theta_{,i}\theta_{,j} \geq 0, \tag{4.86}$$

which implies that μ_{ijkl} and κ_{ij} are positive definite.

4.5 Stokesian Fluids

For fluids, we begin with the spatial forms of the energy equation in Eq. (3.103) and the C–D inequality in Eq. (3.104):

$$\rho(\dot{\chi} + \eta\dot{\theta} + \dot{\eta}\theta) = \tau_{ij} v_{j,i} + \rho r - q_{i,i}, \tag{4.87}$$

$$-\rho(\dot{\chi} + \eta\dot{\theta}) + \tau_{ij} v_{j,i} - \frac{q_i \theta_{,i}}{\theta} \geq 0. \tag{4.88}$$

Consider the possibility of

$$\begin{aligned} \chi &= \chi(\rho^{-1}; d_{kl}; \theta; \theta_{,k}), \\ \tau_{kl} &= \tau_{kl}(\rho^{-1}; d_{kl}; \theta; \theta_{,k}), \\ q_k &= q_k(\rho^{-1}; d_{kl}; \theta; \theta_{,k}). \end{aligned} \tag{4.89}$$

We note that the arguments of the constitutive functions in Eq. (4.89) are objective tensors of various orders. The substitution of Eq. $(4.89)_1$ into Eq. (4.88) gives

$$\begin{aligned} &- \rho \frac{\partial \chi}{\partial (\rho^{-1})} \frac{d\rho^{-1}}{dt} - \rho \frac{\partial \chi}{\partial d_{kl}} \dot{d}_{kl} - \rho \frac{\partial \chi}{\partial \theta} \dot{\theta} - \rho \frac{\partial \chi}{\partial \theta_{,k}} \frac{d\theta_{,k}}{dt} \\ &- \rho\eta\dot{\theta} + \tau_{ij} d_{ij} - \frac{q_i \theta_{,i}}{\theta} \geq 0. \end{aligned} \tag{4.90}$$

The first term in Eq. (4.90) may be written as

$$\begin{aligned} - \rho \frac{\partial \chi}{\partial (\rho^{-1})} \frac{d\rho^{-1}}{dt} &= -\rho \frac{\partial \chi}{\partial (\rho^{-1})} (-1)\rho^{-2} \frac{d\rho}{dt} \\ &= \frac{\partial \chi}{\partial (\rho^{-1})} \rho^{-1}(-\rho v_{k,k}) = -\frac{\partial \chi}{\partial (\rho^{-1})} \delta_{ij} d_{ij} = p\delta_{ij} d_{ij}, \end{aligned} \tag{4.91}$$

where we have used the conservation of mass and introduced the thermodynamic pressure p through

$$p = -\frac{\partial \chi}{\partial \left(\rho^{-1}\right)}. \tag{4.92}$$

Substituting Eq. (4.91) into Eq. (4.90), we have

$$\begin{aligned} &-\rho\frac{\partial \chi}{\partial d_{kl}}\dot{d}_{kl} - \rho\left(\frac{\partial \chi}{\partial \theta} + \eta\right)\dot{\theta} - \rho\frac{\partial \chi}{\partial \theta_{,k}}\frac{d\theta_{,k}}{dt} \\ &+ (\tau_{ij} + p\delta_{ij})d_{ij} - \frac{q_i\theta_{,i}}{\theta} \geq 0. \end{aligned} \tag{4.93}$$

Equation (4.93) is linear in $\dot{d}_{kl}$, $\dot{\theta}$ and $d\theta_{,k}/dt$. For the equation to be true in general, we must have

$$\frac{\partial \chi}{\partial d_{kl}} = 0, \quad \frac{\partial \chi}{\partial \theta} + \eta = 0, \quad \frac{\partial \chi}{\partial \theta_{,k}} = 0. \tag{4.94}$$

Hence,

$$\chi = \chi\left(\rho^{-1}; \theta\right), \tag{4.95}$$

and Eq. (4.93) reduces to

$$(\tau_{ij} + p\delta_{ij})d_{ij} - \frac{q_i\theta_{,i}}{\theta} \geq 0. \tag{4.96}$$

Let

$$\begin{aligned} &\tau_{kl} + p\delta_{kl} = \tau_{kl}^D, \\ &\text{or} \quad \tau_{kl} = -p\delta_{kl} + \tau_{kl}^D, \end{aligned} \tag{4.97}$$

where τ^D is the dissipative or viscous stress. Then Eq. (4.96) takes the following form:

$$\tau_{kl}^D d_{kl} - \frac{q_i\theta_{,i}}{\theta} \geq 0. \tag{4.98}$$

In summary, the free energy has the form of

$$\chi = \chi(\rho^{-1}; \theta). \tag{4.99}$$

The kinetic equation of state is given by

$$p = -\frac{\partial \chi}{\partial(\rho^{-1})}. \tag{4.100}$$

The caloric equation of state is

$$\eta = -\frac{\partial \chi}{\partial \theta}. \tag{4.101}$$

The dissipative constitutive relations may have the following forms:

$$\tau_{kl}^{D} = \tau_{kl}^{D}(\rho^{-1}; d_{kl};\ \theta; \theta_{,k}), \tag{4.102}$$

$$q_k = q_k(\rho^{-1}; d_{kl}; \theta; \theta_{,k}). \tag{4.103}$$

Equations (4.102) and (4.103) are restricted by Eq. (4.98) and the principle of objectivity. At the same time, Eq. (4.87) becomes

$$\rho\theta\frac{d\eta}{dt} = \tau_{kl}^{D} d_{kl} + \rho r - q_{j,j}. \tag{4.104}$$

4.6 Newtonian Fluids

For fluids in the spatial description, the principle of objectivity implies that fluids are spatially isotropic. Newtonian fluids are linear mechanically. The viscous stress and the deformation rate are related by a fourth-order tensor. Since a symmetric and isotropic fourth-order tensor has two parameters only, we have [5]

$$\tau_{ij}^{D} = \lambda v_{k,k}\delta_{ij} + \mu(v_{j,i} + v_{i,j}), \tag{4.105}$$

where λ and μ are the viscosity coefficients for fluids. Then

$$\tau_{ij} = -p\delta_{ij} + \lambda v_{k,k}\delta_{ij} + \mu(v_{j,i} + v_{i,j}). \tag{4.106}$$

We define the mechanical pressure as the negative of the mean normal stress as [15]

$$p^{m} = -\frac{1}{3}\tau_{ii}. \tag{4.107}$$

Setting $j = i$ in Eq. (4.106), we obtain

$$p - p^{m} = \left(\lambda + \frac{2}{3}\mu\right) v_{k,k} = -\left(\lambda + \frac{2}{3}\mu\right)\frac{1}{\rho}\frac{d\rho}{dt} = \kappa v_{k,k}, \tag{4.108}$$

where the conservation of mass has been used and

$$\kappa = \lambda + \frac{2}{3}\mu \tag{4.109}$$

is the coefficient of bulk viscosity for fluids. If we assume Stokes' condition [15]

$$\lambda = -\frac{2}{3}\mu, \tag{4.110}$$

then $\kappa = 0$ and Eq. (4.108) reduces to

$$p = p^m. \tag{4.111}$$

In this case, Eq. (4.105) becomes

$$\tau_{ij}^D = -\frac{2}{3}\mu v_{k,k}\delta_{ij} + \mu(v_{j,i} + v_{i,j}), \tag{4.112}$$

which is deviatoric, i.e., $\tau_{kk}^D = 0$. In addition,

$$q_k = -\kappa_{kl}\theta_{,l}, \tag{4.113}$$

$$\rho\theta\dot{\eta} = \tau_{ij}^D d_{ij} + \rho r - q_{i,i}. \tag{4.114}$$

Some of the above equations can be further linearized thermally by expansions at a reference temperature T_0.

4.7 Ideal Fluids

For ideal fluids, from the energy equation in Eq. (3.108), we have

$$\rho\dot{\varepsilon} = \tau_{mk}v_{k,m} + \rho r - q_{i,i}. \tag{4.115}$$

Consider adiabatic processes with $\mathbf{q} = 0$ and $r = 0$. Equation (4.115) reduces to

$$\rho\dot{\varepsilon} = \tau_{ij}v_{j,i}. \tag{4.116}$$

The energy density is taken as [5]

$$\varepsilon = \varepsilon(\rho^{-1}). \tag{4.117}$$

Then

$$\begin{aligned}\rho\dot{\varepsilon} &= \rho\frac{\partial\varepsilon}{\partial(\rho^{-1})}\frac{d(\rho^{-1})}{dt} = \rho\frac{\partial\varepsilon}{\partial(\rho^{-1})}\left(-\rho^{-2}\frac{d\rho}{dt}\right) \\ &= -\rho^{-1}\frac{\partial\varepsilon}{\partial(\rho^{-1})}(-\rho v_{j,j}) = \frac{\partial\varepsilon}{\partial(\rho^{-1})}v_{j,j} = -p\delta_{ij}v_{j,i},\end{aligned} \tag{4.118}$$

where we have used the conservation of mass and denoted the thermodynamic pressure by

$$p = -\frac{\partial\varepsilon}{\partial(\rho^{-1})}. \tag{4.119}$$

Substituting Eq. (4.118) into the energy equation in Eq. (4.116), we obtain

$$-p\delta_{ij}v_{j,i} = \tau_{ij}v_{j,i}, \tag{4.120}$$

or

$$(\tau_{ij} + p\delta_{ij})v_{j,i} = 0. \tag{4.121}$$

Hence, the stress is given by

$$\tau_{ij} = -p\delta_{ij}. \tag{4.122}$$

Since the motion is adiabatic, we have [16]

$$\frac{d\eta}{dt} = \frac{\partial\eta}{\partial t} + v_j\eta_{,j} = 0. \tag{4.123}$$

With the conservation of mass

$$\frac{\partial\rho}{\partial t} + (\rho v_j)_{,j} = 0, \tag{4.124}$$

the adiabatic condition can be written as

$$\frac{\partial(\rho\eta)}{\partial t} + (\rho\eta v_j)_{,j} = 0, \tag{4.125}$$

where $\rho\eta\mathbf{v}$ is the entropy flux density.

4.8 State Equations

Consider the kinetic equation of state in Eq. (4.100):

$$p = p(\rho, \theta). \tag{4.126}$$

Some common examples are [5]

$$\begin{aligned} & p = p(\rho), \quad \text{barotropic}, \\ & p = p_0 \left(\frac{\rho}{\rho_0}\right)^\gamma, \quad \text{polytropic}, \\ & p = R\rho\theta, \quad \text{ideal gas}, \\ & \rho = \text{constant}, \quad \text{incompressible fluid}. \end{aligned} \tag{4.127}$$

For barotropic fluids, surfaces of constant pressure and constant density coincide with each other.

Consider an ideal gas in a reversible process [1]. From thermodynamics,

$$\begin{aligned} d\eta &= \frac{\Delta Q}{\theta} = \frac{1}{\theta}\left(d\varepsilon + pd\frac{1}{\rho}\right) \\ &= \frac{1}{\theta}\frac{\partial \varepsilon}{\partial \theta}d\theta + \left(\frac{1}{\theta}\frac{\partial \varepsilon}{\partial \rho} - \frac{R}{\rho}\right)d\rho, \end{aligned} \tag{4.128}$$

where

$$\varepsilon = \varepsilon(\rho, \theta), \quad p = R\rho\theta \tag{4.129}$$

have been used. Equation (4.128) is an exact differential. Therefore,

$$\frac{\partial}{\partial \rho}\left(\frac{1}{\theta}\frac{\partial \varepsilon}{\partial \theta}\right) = \frac{\partial}{\partial \theta}\left(\frac{1}{\theta}\frac{\partial \varepsilon}{\partial \rho} - \frac{R}{\rho}\right), \tag{4.130}$$

or

$$\frac{1}{\theta}\frac{\partial^2 \varepsilon}{\partial \theta \partial \rho} = \frac{1}{\theta}\frac{\partial^2 \varepsilon}{\partial \rho \partial \theta} - \frac{1}{\theta^2}\frac{\partial \varepsilon}{\partial \rho}, \tag{4.131}$$

$$0 = \frac{1}{\theta^2}\frac{\partial \varepsilon}{\partial \rho}. \tag{4.132}$$

Hence,

$$\varepsilon = \varepsilon(\theta), \tag{4.133}$$

i.e., the internal energy density of an ideal gas is a function of temperature only. Since

$$\Delta Q = d\varepsilon + pdV, \quad V = \frac{1}{\rho}, \tag{4.134}$$

the specific heat under a constant V is

$$c_V = \left.\frac{\Delta Q}{\Delta\theta}\right|_V = \frac{d\varepsilon}{d\theta}. \tag{4.135}$$

In the case of a constant c_V, by integrating Eq. (4.135), we have

$$\varepsilon = c_V\theta. \tag{4.136}$$

The specific heat under a constant p is

$$c_p = \left.\frac{\Delta Q}{\Delta\theta}\right|_p = \frac{d\varepsilon}{d\theta} + p\left.\frac{dV}{d\theta}\right|_p = c_V + R. \tag{4.137}$$

Denote

$$\gamma = \frac{c_p}{c_V}. \tag{4.138}$$

Then

$$\frac{R}{c_V} = \frac{c_p - c_V}{c_V} = \gamma - 1. \tag{4.139}$$

Substituting Eq. (4.136) into the following from Eq. (4.128):

$$d\eta = \frac{1}{\theta}\left(d\varepsilon + pd\frac{1}{\rho}\right), \tag{4.140}$$

we have

$$\begin{aligned} d\eta &= \frac{c_V d\theta}{\theta} - \frac{R}{\rho}d\rho \\ &= c_V\left[d\ln\theta - (\gamma - 1)d\ln\rho\right], \end{aligned} \tag{4.141}$$

where Eq. (4.139) has been used. Integrating Eq. (4.141), we obtain

$$\eta = c_V \ln \frac{\theta}{\rho^{\gamma-1}} + \text{constant}. \tag{4.142}$$

After the use of

$$\theta = \frac{p}{\rho R}, \tag{4.143}$$

Eq. (4.142) becomes

$$\eta = c_V \ln \frac{p}{\rho^{\gamma}} + \text{constant}. \tag{4.144}$$

Problems

4.1. Differentiate the strain energy density in Eq. (4.23) to obtain the second Piola–Kirchhoff stress in Eq. (4.24).
4.2. Show Eq. (4.31).
4.3. Obtain the constitutive relation of a nonlinear elastic material that is incompressible.
4.4. From the positive definiteness of the strain energy density, show that $E > 0$ and $-1 < \nu < 1/2$.
4.5. Invert the following constitutive relation for a linear elastic material to obtain an expression of strain in terms of stress:

$$\sigma_{ij} = \lambda \varepsilon_{kk} \delta_{ij} + 2\mu \varepsilon_{ij}.$$

4.6. Invert the following constitutive relation for a linear elastic material to obtain an expression of stress in terms of strain:

$$\varepsilon_{ij} = \frac{1+\nu}{E} \sigma_{ij} - \frac{\nu}{E} \sigma_{kk} \delta_{ij}.$$

4.7. Write Eq. (4.54) in the matrix form of Eq. (4.51).
4.8. For a second-order symmetric tensor, establish relationships between the I_1, I_2, I_3 in Eq. (4.30) and the J_1, J_2, J_3 in Eq. (3.65).
4.9. Study the implications of Eq. (4.98) on the μ of a Newtonian fluid satisfying Stokes' assumption.

4.10. For an incompressible Newtonian fluid with

$$\tau_{ij}^{D} = 2\mu d_{ij}, \quad q_k = -\kappa_{kl}\theta_{,l}, \quad \varepsilon = c\theta,$$

show that the energy equation takes the following form:

$$\rho c \frac{d\theta}{dt} = 2\mu d_{ij} d_{ij} + \rho r + \kappa_{ij}\theta_{,ij}.$$

4.11. For barotropic fluids, show that surfaces of constant pressure and constant density coincide with each other.

4.12. Study the implications of the invariance under Eq. (2.106) (principle of objectivity) on Eqs. (4.102) and (4.103).

Chapter 5

Elastic Solids

This chapter collects the relevant equations in different places of previous chapters and group them together for nonlinear and linear elasticity, respectively, followed by a few solutions of static and dynamic problems of linear elasticity. The governing equations of nonlinear and linear thermoviscoelasticity are gathered in the last two sections.

5.1 Equations for Nonlinear Elasticity

The behavior of a nonlinear elastic material in large deformation is described by its strain energy density in Eq. (4.12):

$$\varepsilon = \varepsilon(E_{KL}), \tag{5.1}$$

where the strain has the following expression from Eq. (2.39):

$$E_{KL} = (y_{i,K}y_{i,L} - \delta_{KL})/2. \tag{5.2}$$

Then the first Piola–Kirchhoff stress is obtained from Eq. (4.21) as

$$K_{Lj} = y_{j,K}\rho^0 \frac{\partial \varepsilon}{\partial E_{KL}}, \tag{5.3}$$

which needs to satisfy the linear momentum equation in Eq. (3.138):

$$K_{Lk,L} + \rho^0 f_k = \rho^0 \ddot{y}_k. \tag{5.4}$$

With successive substitutions, Eq. (5.4) can be written as three equations for the three components of $\mathbf{y}(\mathbf{X},t)$. Once $\mathbf{y}(\mathbf{X},t)$ is obtained, the present mass density is given by the conservation of mass in Eq. (3.8):

$$\rho^0 = \rho J. \tag{5.5}$$

5.2 Equations for Linear Elasticity

The linear theory of elasticity is for small-amplitude motions of an elastic body around its reference state under small mechanical loads. It is assumed that the displacement gradient is infinitesimal. We want to obtain equations linear in $u_{i,M}$ and neglect their products or powers. For small deformation, we have

$$\frac{\partial u_i}{\partial X_K} = \frac{\partial u_i}{\partial y_k} y_{k,K} = \frac{\partial u_i}{\partial y_k}(\delta_{kK} + u_{k,K}) \cong \frac{\partial u_i}{\partial y_k}\delta_{kK}. \tag{5.6}$$

Hence, to the lowest order of approximation, the small displacement gradients calculated from the material and spatial coordinates are approximately equal. The material time derivative of an infinitesimal field $f(\mathbf{y}, t) = f[\mathbf{y}(\mathbf{X}, t), t]$ is approximately equal to:

$$\begin{aligned} \frac{df}{dt} = \left.\frac{\partial f}{\partial t}\right|_{\mathbf{X}\,\text{fixed}} &= \left.\frac{\partial f}{\partial t}\right|_{\mathbf{y}\,\text{fixed}} + \left.\frac{\partial f}{\partial y_i}\right|_{t\,\text{fixed}} \left.\frac{\partial y_i}{\partial t}\right|_{\mathbf{X}\,\text{fixed}} \\ &= \left.\frac{\partial f}{\partial t}\right|_{\mathbf{y}\,\text{fixed}} + \left.\frac{\partial f}{\partial y_i}\right|_{t\,\text{fixed}} v_i \cong \left.\frac{\partial f}{\partial t}\right|_{\mathbf{y}\,\text{fixed}}. \end{aligned} \tag{5.7}$$

In linear elasticity, the capital $\mathbf{X}$ is usually written as the lowercase $\mathbf{x}$, and only lowercase indices are used. Then, using Eq. (4.35), the conservation of mass takes the following form:

$$\begin{aligned} \rho^0 &= \rho J \cong \rho(1 + u_{k,k}), \\ \rho &\cong \frac{\rho^0}{1 + u_{k,k}} \cong \rho^0(1 - u_{k,k}). \end{aligned} \tag{5.8}$$

For later convenience, we summarize the equations of linear elasticity of isotropic materials below. Conservation of mass:

$$\rho^0 = \rho(1 + u_{k,k}). \tag{5.9}$$

Linear momentum equation from Eq. (3.129):

$$\begin{aligned} \sigma_{ij,i} + b_j &= \rho^0 \ddot{u}_j, \\ b_j &= \rho^0 f_j. \end{aligned} \tag{5.10}$$

Angular momentum equation from Eq. (3.130):

$$\sigma_{kl} = \sigma_{lk}. \tag{5.11}$$

Strain-displacement relation from Eq. (2.45):

$$\varepsilon_{ij} = (u_{i,j} + u_{j,i})/2. \tag{5.12}$$

Constitutive relations from Eqs. (4.55) and (4.56):

$$\sigma_{ij} = \lambda \varepsilon_{kk} \delta_{ij} + 2\mu \varepsilon_{ij}, \tag{5.13}$$

$$\varepsilon_{ij} = \frac{1}{E}[(1+\nu)\sigma_{ij} - \nu \sigma_{kk} \delta_{ij}]. \tag{5.14}$$

Compatibility conditions from Eqs. (2.48) and (2.49):

$$\begin{aligned} 2\varepsilon_{23,23} &= \varepsilon_{22,33} + \varepsilon_{33,22}, \\ 2\varepsilon_{31,31} &= \varepsilon_{33,11} + \varepsilon_{11,33}, \\ 2\varepsilon_{12,12} &= \varepsilon_{11,22} + \varepsilon_{22,11}, \end{aligned} \tag{5.15}$$

$$\begin{aligned} \varepsilon_{11,23} &= (\varepsilon_{31,2} + \varepsilon_{12,3} - \varepsilon_{23,1})_{,1}, \\ \varepsilon_{22,31} &= (\varepsilon_{12,3} + \varepsilon_{23,1} - \varepsilon_{31,2})_{,2}, \\ \varepsilon_{33,12} &= (\varepsilon_{23,1} + \varepsilon_{31,2} - \varepsilon_{12,3})_{,3}. \end{aligned} \tag{5.16}$$

In the rest of this chapter, the following unabbreviated notations for the linear strain and stress components may be used:

$$\begin{aligned} \begin{bmatrix} \varepsilon_{11} & \varepsilon_{12} & \varepsilon_{13} \\ \varepsilon_{21} & \varepsilon_{22} & \varepsilon_{23} \\ \varepsilon_{31} & \varepsilon_{32} & \varepsilon_{33} \end{bmatrix} &= \begin{bmatrix} \varepsilon_{x} & \varepsilon_{xy} & \varepsilon_{xz} \\ \varepsilon_{yx} & \varepsilon_{y} & \varepsilon_{yz} \\ \varepsilon_{zx} & \varepsilon_{zy} & \varepsilon_{z} \end{bmatrix}, \\ &= \begin{bmatrix} \varepsilon_{x} & \gamma_{xy}/2 & \gamma_{xz}/2 \\ \gamma_{yx}/2 & \varepsilon_{y} & \gamma_{yz}/2 \\ \gamma_{zx}/2 & \gamma_{zy}/2 & \varepsilon_{z} \end{bmatrix}, \end{aligned} \tag{5.17}$$

$$\begin{bmatrix} \sigma_{11} & \sigma_{12} & \sigma_{13} \\ \sigma_{21} & \sigma_{22} & \sigma_{23} \\ \sigma_{31} & \sigma_{32} & \sigma_{33} \end{bmatrix} = \begin{bmatrix} \sigma_{x} & \tau_{xy} & \tau_{xz} \\ \tau_{yx} & \sigma_{y} & \tau_{yz} \\ \tau_{zx} & \tau_{zy} & \sigma_{z} \end{bmatrix}. \tag{5.18}$$

The equations of linear elasticity can be written as three equations for the three components of the displacement vector. We begin with

$$\begin{aligned}\sigma_{ij} &= \lambda\varepsilon_{kk}\delta_{ij} + 2\mu\varepsilon_{ij}\\ &= \lambda u_{k,k}\delta_{ij} + \mu(u_{i,j} + u_{j,i}).\end{aligned} \tag{5.19}$$

Then

$$\begin{aligned}\sigma_{ij,i} &= \lambda u_{k,ki}\delta_{ij} + \mu(u_{i,ji} + u_{j,ii})\\ &= \lambda u_{k,kj} + \mu(u_{k,kj} + u_{j,kk})\\ &= (\lambda + \mu)u_{k,kj} + \mu u_{j,kk}.\end{aligned} \tag{5.20}$$

Substituting Eq. (5.20) into Eq. (5.10), we obtain Navier's equation as

$$(\lambda + \mu)u_{k,kj} + \mu u_{j,kk} + b_j = \rho^0 \ddot{u}_j, \tag{5.21}$$

or

$$(\lambda + \mu)\nabla(\nabla \cdot \mathbf{u}) + \mu\nabla^2\mathbf{u} + \mathbf{b} = \rho^0\ddot{\mathbf{u}}. \tag{5.22}$$

Similarly, for anisotropic materials, we have

$$c_{ijkl}u_{k,li} + b_j = \rho^0\ddot{u}_j. \tag{5.23}$$

Consider the elastic body in Fig. 5.1. Its boundary surface S is partitioned into two parts:

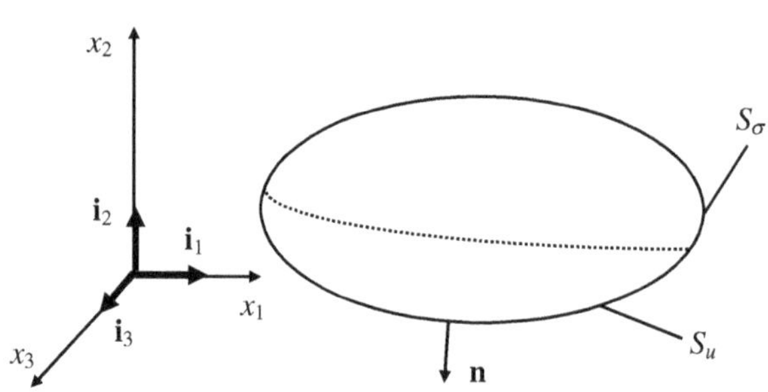

Fig. 5.1. An elastic body and the partition of its boundary surface.

$$S_u \cup S_\sigma = S, \quad S_u \cap S_\sigma = \emptyset. \tag{5.24}$$

The boundary-value problem for elastostatics is

$$\begin{aligned} \sigma_{ji,j} + b_i = 0 \quad & \text{in } V, \\ \sigma_{ij} = c_{ijkl}\varepsilon_{kl} \quad & \text{in } V, \\ \varepsilon_{ij} = (u_{i,j} + u_{j,i})/2 \quad & \text{in } V, \end{aligned} \tag{5.25}$$

and

$$\begin{aligned} u_i = \bar{u}_i \quad & \text{on} \quad S_u, \\ n_j\sigma_{ji} = \bar{t}_i \quad & \text{on} \quad S_\sigma, \end{aligned} \tag{5.26}$$

where $\bar{u}_i$ and $\bar{t}_i$ are prescribed boundary displacement and traction. For dynamic problems, initial conditions are also needed.

5.3 A Layer Under Normal Traction

Consider an unbounded layer with its bottom fixed as shown in Fig. 5.2. Its top is under a uniform normal traction p. The boundary-value problem consists of Navier's equation and boundary conditions:

$$\begin{aligned} & (\lambda + \mu)u_{k,kj} + \mu u_{j,kk} = 0 \quad \text{in } V, \\ & u_j = 0, \quad x_3 = 0, \\ & \sigma_{31} = \sigma_{32} = 0, \quad \sigma_{33} = p, \quad x_3 = h. \end{aligned} \tag{5.27}$$

The trial solution is taken to be

$$u_1 = u_2 = 0, \quad u_3 = u_3(x_3). \tag{5.28}$$

Then the strain field is

$$\varepsilon_{33} = u_{3,3}, \quad \text{all other } \varepsilon_{ij} = 0. \tag{5.29}$$

From

$$\sigma_{ij} = \lambda\varepsilon_{kk}\delta_{ij} + 2\mu\varepsilon_{ij} = \lambda u_{3,3}\delta_{ij} + 2\mu\varepsilon_{ij}, \tag{5.30}$$

the stress field is found to be

$$\begin{aligned} & \sigma_{11} = \sigma_{22} = \lambda u_{3,3}, \quad \sigma_{33} = (\lambda + 2\mu)u_{3,3}, \\ & \text{all other } \sigma_{ij} = 0. \end{aligned} \tag{5.31}$$

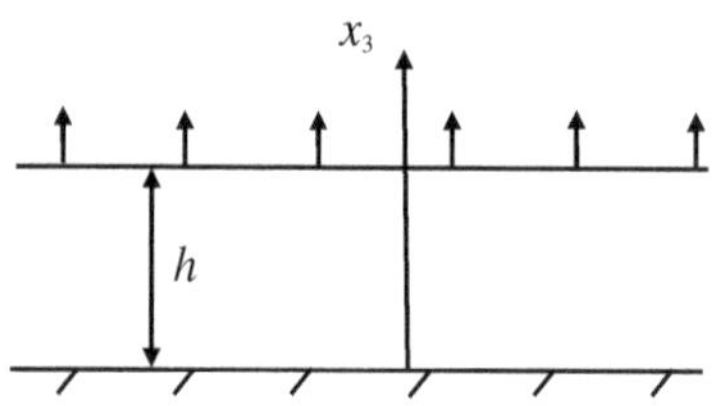

Fig. 5.2. An unbounded layer under a normal traction.

The only nontrivial equilibrium equation is

$$\sigma_{33,3} = 0. \tag{5.32}$$

The boundary-value problem in Eq. (5.27) reduces to

$$\begin{aligned} &(\lambda + 2\mu)u_{3,33} = 0, \quad 0 < x_3 < h, \\ &u_3(0) = 0, \\ &(\lambda + 2\mu)u_{3,3}(h) = p. \end{aligned} \tag{5.33}$$

The general solution for u_3 is

$$u_3 = C_1 x_3 + C_2. \tag{5.34}$$

From the two boundary conditions in Eq. (5.33), we determine

$$C_1 = \frac{p}{\lambda + 2\mu}, \quad C_2 = 0. \tag{5.35}$$

Hence,

$$u_3 = \frac{p}{\lambda + 2\mu} x_3, \tag{5.36}$$

$$\sigma_{33} = p, \quad \sigma_{11} = \sigma_{22} = \frac{\lambda}{\lambda + 2\mu} p. \tag{5.37}$$

We note that in the stress field in Eq. (5.37), σ_{11} and σ_{22} are nonzero.

5.4 A Layer Under Its Own Weight

Consider the unbounded layer in Fig. 5.3. It is under its own weight. The top surface is free. The bottom is fixed. The boundary-value

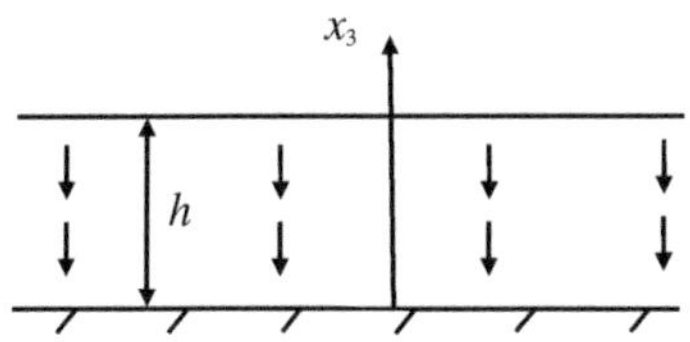

Fig. 5.3. A layer under its own weight.

problem is

$$\begin{aligned} &(\lambda+\mu)u_{k,kj}+\mu u_{j,kk}-\rho^0 g\delta_{j3}=0 \quad \text{in} \quad V,\\ &u_j=0, \quad x_3=0,\\ &\sigma_{3j}=0, \quad x_3=h. \end{aligned} \tag{5.38}$$

The trial solution is

$$u_1=u_2=0, \quad u_3=u_3(x_3). \tag{5.39}$$

The corresponding strain and stress fields are

$$\varepsilon_{33}=u_{3,3}, \quad \text{all other } \varepsilon_{ij}=0, \tag{5.40}$$

and

$$\sigma_{ij}=\lambda\varepsilon_{kk}\delta_{ij}+2\mu\varepsilon_{ij}=\lambda u_{3,3}\delta_{ij}+2\mu\varepsilon_{ij}, \tag{5.41}$$

$$\begin{aligned} \sigma_{11}=\sigma_{22}=\lambda u_{3,3}, \quad \sigma_{33}=(\lambda+2\mu)u_{3,3},\\ \text{all other } \sigma_{ij}=0. \end{aligned} \tag{5.42}$$

The only nontrivial equilibrium equation is

$$\sigma_{33,3}-\rho^0 g=0. \tag{5.43}$$

The boundary-value problem in Eq. (5.38) reduces to

$$\begin{aligned} &(\lambda+2\mu)u_{3,33}=\rho^0 g, \quad 0<x_3<h,\\ &u_3(0)=0,\\ &(\lambda+2\mu)u_{3,3}(h)=0. \end{aligned} \tag{5.44}$$

The general solution for u_3 is

$$u_3=\frac{\rho^0 g}{\lambda+2\mu}\frac{x_3^2}{2}+C_1x_3+C_2. \tag{5.45}$$

From the boundary conditions in Eq. (5.44), we determine

$$C_1 = \frac{-\rho^0 gh}{\lambda + 2\mu}, \quad C_2 = 0. \tag{5.46}$$

Hence,

$$u_3 = \frac{\rho^0 g}{\lambda + 2\mu} \frac{x_3^2 - 2hx_3}{2}, \tag{5.47}$$

and

$$\begin{aligned} \sigma_{33} &= \rho^0 g(x_3 - h), \\ \sigma_{11} = \sigma_{22} &= \frac{\lambda}{\lambda + 2\mu} \rho^0 g(x_3 - h). \end{aligned} \tag{5.48}$$

We note that in the stress field in Eq. (5.48), σ_{11} and σ_{22} are nonzero.

5.5 A Layer Under Shear Traction

Consider the unbounded layer in Fig. 5.4. Its top surface is under a shear traction q. The boundary-value problem is

$$\begin{aligned} &(\lambda + \mu) u_{k,kj} + \mu u_{j,kk} = 0 \quad \text{in} \quad V, \\ &u_j = 0, \quad x_3 = 0, \\ &\sigma_{32} = \sigma_{33} = 0, \quad \sigma_{31} = q, \quad x_3 = h. \end{aligned} \tag{5.49}$$

The trial solution is

$$u_1 = u_1(x_3), \quad u_2 = u_3 = 0. \tag{5.50}$$

The corresponding strain and stress fields are

$$2\varepsilon_{31} = u_{1,3}, \quad \text{all other } \varepsilon_{ij} = 0, \tag{5.51}$$

and

$$\sigma_{31} = \mu u_{1,3}, \quad \text{all other } \sigma_{ij} = 0. \tag{5.52}$$

The only nontrivial equilibrium equation is

$$\sigma_{31,3} = 0. \tag{5.53}$$

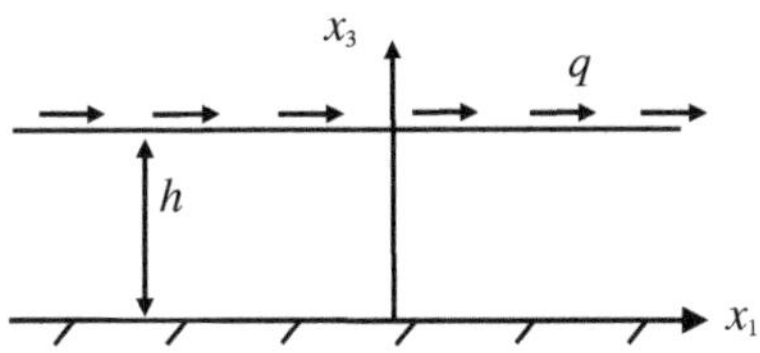

Fig. 5.4. A layer under a shear traction.

The boundary-value problem in Eq. (5.49) reduces to

$$\begin{aligned} &\mu u_{1,33} = 0, \quad 0 < x_3 < h, \\ &u_1(0) = 0, \\ &\mu u_{1,3}(h) = q. \end{aligned} \tag{5.54}$$

The general solution for u_1 is

$$u_1 = C_1 x_3 + C_2. \tag{5.55}$$

From the boundary conditions in Eq. (5.54), we determine

$$C_1 = \frac{q}{\mu}, \quad C_2 = 0. \tag{5.56}$$

Hence,

$$u_1 = \frac{q}{\mu} x_3, \quad \sigma_{31} = q. \tag{5.57}$$

5.6 Two Layers Under Shear Traction

Consider the two perfectly bonded layers in Fig. 5.5. They are under an interface load q_1 per unit interface area and a shear traction q_2 at the top.

Based on the solution of a single layer in the previous section, the trial solution is taken as

$$u_1 = u_1(x_3), \quad u_2 = u_3 = 0. \tag{5.58}$$

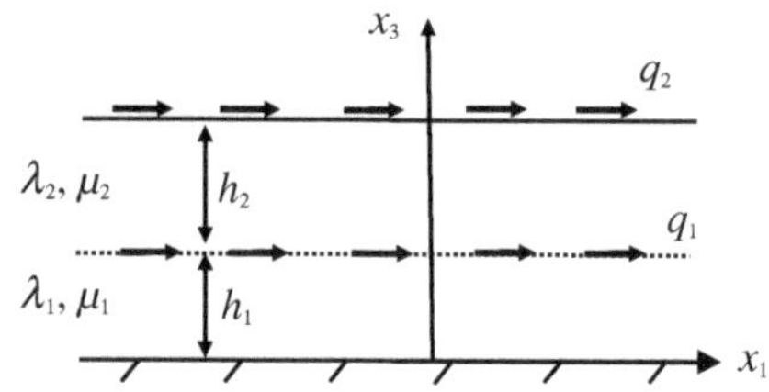

Fig. 5.5. Two perfectly bonded layers under shear traction and interface load.

The strain field is

$$2\varepsilon_{31} = u_{1,3}, \quad \text{all other } \varepsilon_{ij} = 0. \tag{5.59}$$

The material properties of the two layers are different. Hence, the stress field has different expressions in the two layers:

$$\sigma_{31} = \begin{cases} \mu_1 u_{1,3}, & 0 < x_3 < h_1, \\ \mu_2 u_{1,3}, & h_1 < x_3 < h_1 + h_2, \end{cases} \tag{5.60}$$
$$\text{all other } \sigma_{ij} = 0.$$

The only nontrivial equilibrium equation is

$$\sigma_{31,3} = 0. \tag{5.61}$$

The boundary-value problem consists of the following equilibrium equations:

$$\begin{aligned} \mu_1 u_{1,33} &= 0, \quad 0 < x_3 < h_1, \\ \mu_2 u_{1,33} &= 0, \quad h_1 < x_3 < h_1 + h_2, \end{aligned} \tag{5.62}$$

and boundary conditions:

$$\begin{aligned} u_1(0) &= 0, \\ \mu_2 u_{1,3}(h) &= q_2, \end{aligned} \tag{5.63}$$

as well as continuity or jump conditions at the interface:

$$\begin{aligned} u_1(h_1^+) &= u_1(h_1^-), \\ \sigma_{31}(h_1^+) + q_1 &= \sigma_{31}(h_1^-), \end{aligned} \tag{5.64}$$

where Eq. $(5.64)_2$ is obtained by applying Newton's second law to the differential element near the interface in Fig. 5.6.

The general solutions in the two layers are

$$u_1 = \begin{cases} C_1 x_3 + C_2, & 0 < x_3 < h_1, \\ C_3 x_3 + C_4, & h_1 < x_3 < h_1 + h_2, \end{cases} \tag{5.65}$$

where C_1 through C_4 are determined by Eqs. (5.63) and (5.64).

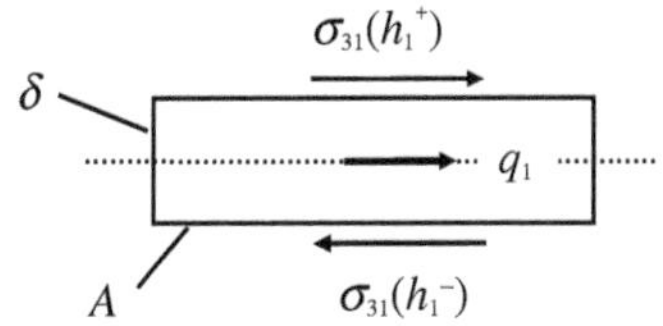

Fig. 5.6. Free-body diagram of a differential element at the interface.

5.7 Plane Waves

Consider plane waves in an unbounded cubic crystal of class (m3m) with three independent elastic constants. In the matrix notation,

$$[c_{pq}] = \begin{bmatrix} c_{11} & c_{12} & c_{12} & 0 & 0 & 0 \\ c_{12} & c_{11} & c_{12} & 0 & 0 & 0 \\ c_{12} & c_{12} & c_{11} & 0 & 0 & 0 \\ 0 & 0 & 0 & c_{44} & 0 & 0 \\ 0 & 0 & 0 & 0 & c_{44} & 0 \\ 0 & 0 & 0 & 0 & 0 & c_{44} \end{bmatrix}. \tag{5.66}$$

The stress-strain relations are

$$\begin{aligned} \sigma_1 &= c_{11}\varepsilon_1 + c_{12}\varepsilon_2 + c_{12}\varepsilon_3, \\ \sigma_2 &= c_{12}\varepsilon_1 + c_{11}\varepsilon_2 + c_{12}\varepsilon_3, \\ \sigma_3 &= c_{12}\varepsilon_1 + c_{12}\varepsilon_2 + c_{11}\varepsilon_3, \\ \sigma_4 &= c_{44}\varepsilon_4, \quad \sigma_5 = c_{44}\varepsilon_5, \quad \sigma_6 = c_{44}\varepsilon_6. \end{aligned} \tag{5.67}$$

Consider plane waves propagating in the x_3 direction without x_1 and x_2 dependence. They are governed by

$$\begin{aligned} c_{44}u_{1,33} &= \rho^0\ddot{u}_1, \\ c_{44}u_{2,33} &= \rho^0\ddot{u}_2, \\ c_{11}u_{3,33} &= \rho^0\ddot{u}_3. \end{aligned} \tag{5.68}$$

In Eq. $(5.68)_1$, let

$$u_1 = U_1 \exp[i(\zeta x_3 - \omega t)], \tag{5.69}$$

which is a transverse or shear wave because the displacement u_1 is perpendicular to the propagation direction x_3. The substitution of

Eq. (5.69) into Eq. $(5.68)_1$ leads to the following wave speed:

$$v = \frac{\omega}{\zeta} = \sqrt{\frac{c_{44}}{\rho^0}}. \tag{5.70}$$

In the special case of isotropic materials, Eq. (5.70) reduces to

$$c_{44} = \mu, \quad v = \sqrt{\frac{\mu}{\rho^0}} = c_2. \tag{5.71}$$

Equation $(5.68)_2$ also describes a shear wave with the same speed. For Eq. $(5.68)_3$, let

$$u_3 = U_3 \exp[i(\zeta x_3 - \omega t)], \tag{5.72}$$

which is a longitudinal wave because the displacement u_3 is parallel to the propagation direction x_3. The substitution of Eq. (5.72) into Eq. $(5.68)_3$ leads to the following wave speed:

$$v = \frac{\omega}{\zeta} = \sqrt{\frac{c_{11}}{\rho^0}}. \tag{5.73}$$

For an isotropic material,

$$c_{11} = \lambda + 2\mu, \quad v = \sqrt{\frac{\lambda + 2\mu}{\rho^0}} = c_1. \tag{5.74}$$

Consider waves propagating in anisotropic elastic materials in general governed by Eq. (5.23) with $\mathbf{b} = 0$:

$$c_{ijkl} u_{k,li} = \rho^0 \ddot{u}_j. \tag{5.75}$$

We are interested in the following plane wave:

$$u_k = A_k f(\mathbf{n} \cdot \mathbf{x} - vt) \tag{5.76}$$

where $\mathbf{A}$, $\mathbf{n}$ and v are constants. f is an arbitrary function. $\mathbf{n} \cdot \mathbf{x} - vt$ is the phase of the wave. v is the phase velocity. $\mathbf{n} \cdot \mathbf{x} - vt =$ constant determines the wave front, which in this case is a plane with a normal $\mathbf{n}$. Differentiating Eq. (5.76), we obtain

$$u_{k,l} = A_k f' n_l, \quad u_{k,li} = A_k f'' n_l n_i, \quad \ddot{u}_k = A_k f'' v^2. \tag{5.77}$$

The substitution of Eq. (5.77) into Eq. (5.75) yields the following linear homogeneous equation for **A**:

$$c_{ijkl}A_k n_l n_i = \rho^0 v^2 A_j, \tag{5.78}$$

which can be written as

$$(\Gamma_{jk} - \rho^0 v^2 \delta_{jk})A_k = 0, \tag{5.79}$$

where we have denoted

$$\Gamma_{jk} = c_{ijkl}n_l n_i = \Gamma_{kj}. \tag{5.80}$$

Γ_{jk} is the acoustic tensor or Christoffel tensor. Equation (5.79) is an eigenvalue problem of the acoustic tensor. For nontrivial solutions of **A**, we must have

$$\det(\Gamma_{jk} - \rho^0 v^2 \delta_{jk}) = 0, \tag{5.81}$$

which is a polynomial equation of degree three for v^2. Since the acoustic tensor is assumed to be real, symmetric and positive definite, there exist three real and positive eigenvalues which lead to three wave speeds:

$$v^{(1)}, v^{(2)}, v^{(3)}, \tag{5.82}$$

and three corresponding eigenvectors that are orthogonal:

$$A_j^{(1)}, A_j^{(2)}, A_j^{(3)}. \tag{5.83}$$

Graphically, given a propagation direction **n**, there exist three plane waves with velocities $v^{(1)}$, $v^{(2)}$ and $v^{(3)}$. Their displacement vectors are perpendicular to each other as shown in Fig. 5.7. In anisotropic materials, typically one of the waves, e.g., the one with $\mathbf{u}^{(1)}$, is roughly aligned with **n**. This is called a quasi-longitudinal wave. The other two waves with $\mathbf{u}^{(2)}$ and $\mathbf{u}^{(3)}$ have their displacement vectors roughly perpendicular to **n** and are called quasi-transverse waves. In materials with high symmetries, e.g. isotropic materials, one wave is exactly longitudinal, and the other two are exactly transverse.

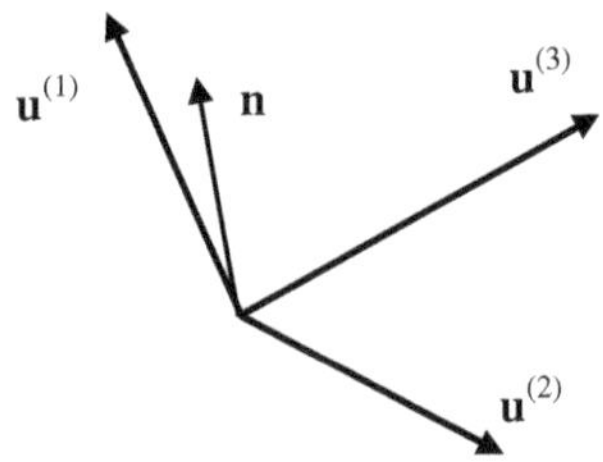

Fig. 5.7. Plane waves propagating in the direction of $\mathbf{n}$.

5.8 Reflection

Consider a so-called antiplane or shear horizontal (SH) wave with $u_3 = u(x_1, x_2, t)$ alone ($u_1 = u_2 = 0$) incident upon the plane boundary of a half-space as shown in Fig. 5.8. The boundary is traction-free. In a semi-infinite medium, a wave solution needs to satisfy boundary conditions in addition to Navier's equation. Specifically, the incident and reflected waves together must satisfy the following equation and boundary condition:

$$\begin{aligned} &\mu\nabla^2 u = \rho^0 \ddot{u}, \quad x_2 > 0, \\ &\sigma_{23} = 0, \quad x_2 = 0. \end{aligned} \tag{5.84}$$

The incident wave can be written as

$$u = A\exp[i\xi(x_1 \sin\alpha - x_2\cos\alpha - vt)], \tag{5.85}$$

which is considered known. The incident wave itself satisfies the differential equations in Eq. $(5.84)_1$ provided that

$$\mu(-\xi^2\sin^2\alpha - \xi^2\cos^2\alpha) = \rho^0(-\xi^2 v^2), \tag{5.86}$$

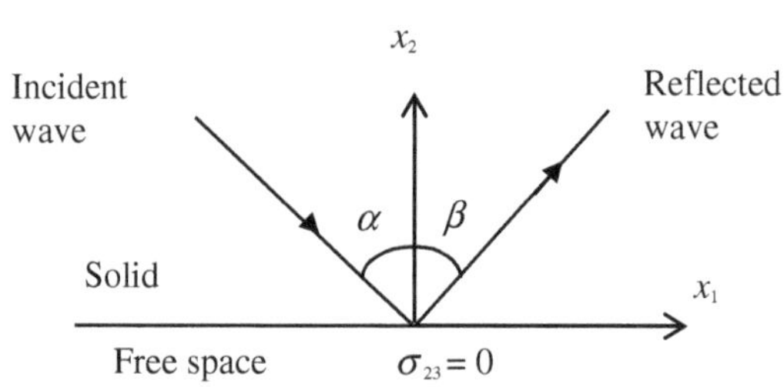

Fig. 5.8. Incident and reflected waves at the plane boundary of an elastic half-space.

or

$$v^2 = \frac{\mu}{\rho^0} = v_T^2 = c_2^2, \tag{5.87}$$

where v_T (or c_2 in the case of isotropic materials) is the speed of transverse plane waves. The stress component needed for the boundary condition is

$$\sigma_{23} = \mu u_{,2} = \mu(-i\xi \cos\alpha) A \exp[i\xi(x_1 \sin\alpha - x_2 \cos\alpha - vt)]. \tag{5.88}$$

Similarly, we write the reflected wave as

$$u = B \exp[i\xi(x_1 \sin\beta + x_2 \cos\beta - vt)], \tag{5.89}$$

which also satisfies Eq. $(5.84)_1$ and is considered unknown. For the boundary condition, we need

$$\sigma_{23} = \mu(i\xi \cos\beta) B \exp[i\xi(x_1 \sin\beta + x_2 \cos\beta - vt)]. \tag{5.90}$$

The incident and reflected waves already satisfy the governing equation individually, and so does their sum. At the boundary, the sum of the incident and reflected waves together has to satisfy Eq. $(5.84)_2$, i.e.,

$$\begin{aligned} &\mu(-i\xi \cos\alpha) A \exp[i\xi(x_1 \sin\alpha - vt)] \\ &\quad + \mu(i\xi \cos\beta) B \exp[i\xi(x_1 \sin\beta - vt)] = 0, \end{aligned} \tag{5.91}$$

for any x_1 and any t. This implies that

$$\beta = \alpha, \quad B = A, \tag{5.92}$$

and thus determines the reflected wave.

5.9 Reflection and Refraction

Consider two half-spaces of different elastic materials as shown in Fig. 5.9. The motion is antiplane with $u_3 = u(x_1, x_2, t)$ and $u_1 =$

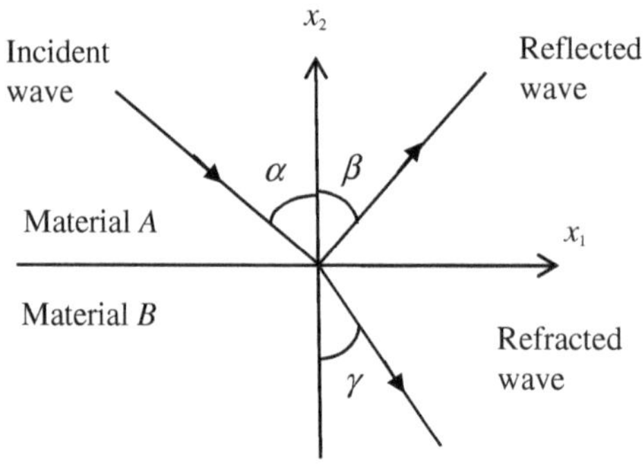

Fig. 5.9. Incident, reflected, and refracted waves at a plane interface between two elastic half-spaces.

$u_2 = 0$. The incident and reflected waves together must satisfy the following equation in material A:

$$\mu_A \nabla^2 u = \rho_A^0 \ddot{u}, \quad x_2 > 0. \tag{5.93}$$

The refracted wave must satisfy the following equation in material B:

$$\mu_B \nabla^2 u = \rho_B^0 \ddot{u}, \quad x_2 < 0. \tag{5.94}$$

The interface continuity conditions are

$$\begin{aligned} u(x_2 = 0^+) &= u(x_2 = 0^-), \\ \sigma_{23}(x_2 = 0^+) &= \sigma_{23}(x_2 = 0^-). \end{aligned} \tag{5.95}$$

The incident wave can be written as

$$\begin{aligned} u &= A \exp[i\xi_A(x_1 \sin\alpha - x_2 \cos\alpha - v_A t)], \\ v_A^2 &= \frac{\mu_A}{\rho_A^0}, \\ \sigma_{23} &= \mu_A(-i\xi_A \cos\alpha) A \exp[i\xi_A(x_1 \sin\alpha - x_2 \cos\alpha - v_A t)], \end{aligned} \tag{5.96}$$

which is considered known. The reflected wave can be written as

$$\begin{aligned} u &= B \exp[i\xi_A(x_1 \sin\beta + x_2 \cos\beta - v_A t)], \\ \sigma_{23} &= \mu_A(i\xi_A \cos\beta) B \exp[i\xi_A(x_1 \sin\beta + x_2 \cos\beta - v_A t)], \end{aligned} \tag{5.97}$$

which is considered unknown. We write the refracted wave as

$$\begin{aligned}u &= C\exp[i\xi_B(x_1\sin\gamma - x_2\cos\gamma - v_B t)],\\ v_B^2 &= \frac{\mu_B}{\rho_B^0},\\ \sigma_{23} &= \mu_B(-i\xi_B\cos\gamma)C\exp[i\xi_B(x_1\sin\gamma - x_2\cos\gamma - v_B t)].\end{aligned} \tag{5.98}$$

Equations (5.93) and (5.94) are already satisfied. From (5.95), we have

$$\begin{aligned}&A\exp[i\xi_A(x_1\sin\alpha - v_A t)] + B\exp[i\xi_A(x_1\sin\beta - v_A t)]\\ &\quad = C\exp[i\xi_B(x_1\sin\gamma - v_B t)],\\ &\mu_A(-i\xi_A\cos\alpha)A\exp[i\xi_A(x_1\sin\alpha - v_A t)]\\ &\qquad + \mu_A(i\xi_A\cos\beta)B\exp[i\xi_A(x_1\sin\beta - v_A t)]\\ &\quad = \mu_B(-i\xi_B\cos\gamma)C\exp[i\xi_B(x_1\sin\gamma - v_B t)].\end{aligned} \tag{5.99}$$

Equation (5.99) can be satisfied if the following two sets of relations are true:

$$\begin{aligned}&\exp[i\xi_A(x_1\sin\alpha - v_A t)]\\ &\quad = \exp[i\xi_A(x_1\sin\beta - v_A t)]\\ &\quad = \exp[i\xi_B(x_1\sin\gamma - v_B t)],\end{aligned} \tag{5.100}$$

and

$$\begin{aligned}A + B &= C,\\ \mu_A\xi_A A\cos\alpha - \mu_A\xi_A B\cos\beta &= \mu_B\xi_B C\cos\gamma.\end{aligned} \tag{5.101}$$

We analyze Eqs. (5.100) and (5.101) separately. For Eq. (5.100) to be true for any t and any x_1, we must have

$$\begin{aligned}&\xi_A v_A = \xi_B v_B,\\ &\xi_A\sin\alpha = \xi_A\sin\beta = \xi_B\sin\gamma.\end{aligned} \tag{5.102}$$

Hence,

$$\frac{\xi_B}{\xi_A} = \frac{v_A}{v_B} = \sqrt{\frac{\mu_A\rho_B^0}{\rho_A^0\mu_B}} = \frac{1}{n}, \tag{5.103}$$

where n is the refractive index, and

$$\beta = \alpha, \qquad \frac{\sin\gamma}{\sin\alpha} = \frac{\xi_A}{\xi_B} = \frac{v_B}{v_A} = n, \tag{5.104}$$

which is Snell's law. Then, from Eq. (5.101), we solve for B and C with the following results:

$$\begin{aligned} \frac{B}{A} &= \frac{\rho_A^0 \sin 2\alpha - \rho_B^0 \sin 2\gamma}{\rho_A^0 \sin 2\alpha + \rho_B^0 \sin 2\gamma}, \\ \frac{C}{A} &= \frac{2\rho_A^0 \sin 2\alpha}{\rho_A^0 \sin 2\alpha + \rho_B^0 \sin 2\gamma}. \end{aligned} \tag{5.105}$$

Thus the reflected and refracted waves are fully determined. As a special case, consider normal incidence with $\alpha = 0$. From Eq. (5.104), we have $\beta = \gamma = 0$. In this case Eq. (5.105) reduces to

$$\begin{aligned} \frac{B}{A} &= \frac{\rho_A^0 - n\rho_B^0}{\rho_A^0 + n\rho_B^0} = \frac{\rho_A^0 v_A - \rho_B^0 v_B}{\rho_A^0 v_A + \rho_B^0 v_B}, \\ \frac{C}{A} &= \frac{2\rho_A^0 v_A}{\rho_A^0 v_A + \rho_B^0 v_B}, \end{aligned} \tag{5.106}$$

where $\rho_A^0 v_A$ or $\rho_B^0 v_B$ is the acoustic impedance. When

$$\rho_A^0 v_A = \rho_B^0 v_B, \tag{5.107}$$

Equation (5.106) reduces to

$$B = 0, \qquad C = 1. \tag{5.108}$$

In this case, the wave is not reflected and total transmission occurs.

5.10 Thickness Vibrations of Plates

Consider a plate of cubic crystals as shown in Fig. 5.10. It has a uniform thickness $2h$ and is unbounded in the x_1 and x_2 directions. The two surfaces are traction-free. The boundary conditions are

$$\sigma_{3j} = 0, \qquad x_3 = \pm h. \tag{5.109}$$

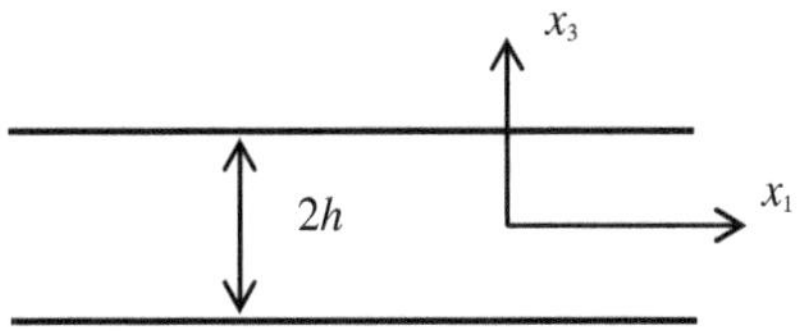

Fig. 5.10. Cubic crystal plate with $x_1 - x_3$ coordinates.

Thickness modes depend on the plate thickness coordinate x_3 and time only. For cubic crystals, thickness vibrations separate into thickness-shear and thickness-stretch modes. Consider thickness-shear modes described by $u_1(x_3, t)$ first. They are governed by

$$\begin{aligned} c_{44}u_{1,33} &= \rho^0 \ddot{u}_1, \quad -h < x_3 < h, \\ \sigma_{31} &= c_{44}u_{1,3} = 0, \quad x_3 = \pm h. \end{aligned} \tag{5.110}$$

Consider modes antisymmetric about the plate middle plane first. Let

$$u_1 = U_1 \sin \zeta x_3 \exp(i\omega t). \tag{5.111}$$

The boundary conditions in Eq. $(5.110)_2$ imply that

$$\zeta^{(n)} = \frac{n\pi}{2h}, \quad n = 1, 3, 5, \ldots. \tag{5.112}$$

Equation $(5.110)_1$ requires that

$$\omega^{(n)} = \sqrt{\frac{c_{44}}{\rho^0}}\zeta^{(n)} = \sqrt{\frac{c_{44}}{\rho^0}}\frac{n\pi}{2h}, \quad n = 1, 3, 5, \ldots. \tag{5.113}$$

$n = 1$ gives the fundamental thickness-shear frequency and mode. $n > 1$ are the overtone frequencies and modes. Equation (5.113) shows that the overtones are integral multiples of the fundamental and are called harmonic overtones or harmonics. If $\cos \zeta x_3$ is used in Eq. (5.111), a different set of modes symmetric about $x_3 = 0$ with $n = 0, 2, 4, \ldots$ will be obtained. $n = 0$ represents a rigid-body mode and is not of interest here. Thickness-shear modes described by u_2 are similar.

Thickness-stretch vibrations described by $u_3(x_3, t)$ are governed by

$$\begin{aligned} &c_{11}u_{3,33} = \rho^0 \ddot{u}_3, \quad -h < x_3 < h, \\ &\sigma_{33} = c_{11}u_{3,3} = 0, \quad x_3 = \pm h. \end{aligned} \tag{5.114}$$

Let

$$u_3 = U_3 \sin \zeta x_3 \exp(i\omega t). \tag{5.115}$$

The boundary conditions in Eq. $(5.114)_2$ imply that

$$\zeta^{(n)} = \frac{n\pi}{2h}, \quad n = 1, 3, 5, \ldots. \tag{5.116}$$

Equation $(5.114)_1$ requires that

$$\omega^{(n)} = \sqrt{\frac{c_{11}}{\rho^0}}\zeta^{(n)} = \sqrt{\frac{c_{11}}{\rho^0}}\frac{n\pi}{2h}, \quad n = 1, 3, 5, \ldots. \tag{5.117}$$

If $\cos \zeta x_3$ is used in Eq. (5.115), a different set of modes with $n = 0, 2, 4, \ldots$ will be obtained.

5.11 Propagating Waves in Plates

Consider antiplane waves described by $u_1 = 0$, $u_2 = 0$ and $u_3 = u_3(x_1, x_2, t)$ in a plate of cubic crystals as shown in Fig. 5.11. The surfaces of the plate are traction free.

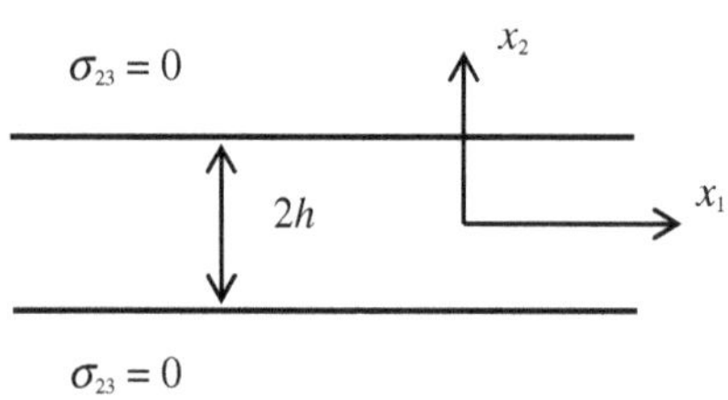

Fig. 5.11. Cubic crystal plate with $x_1 - x_2$ coordinates and $\sigma_{23} = 0$ boundaries.

The governing equation is

$$\begin{aligned} \sigma_{13,1} + \sigma_{23,2} &= c_{44}(u_{3,11} + u_{3,22}) \\ &= c_{44}\nabla^2 u_3 = \rho^0 \ddot{u}_3, \quad -h < x_2 < h. \end{aligned} \tag{5.118}$$

The boundary conditions are

$$\sigma_{23} = c_{44} u_{3,2} = 0, \quad x_2 = \pm h. \tag{5.119}$$

The waves may be symmetric or antisymmetric about the plate middle plane. They are discussed separately below. For antisymmetric waves, we consider the possibility of

$$u_3 = U_3 \sin \eta x_2 \cos(\xi x_1 - \omega t). \tag{5.120}$$

For Eq. (5.120) to satisfy Eq. (5.118), we must have

$$\begin{aligned} \eta^2 &= \frac{\rho^0 \omega^2}{c_{44}} - \xi^2 = \xi^2 \left(\frac{v^2}{v_T^2} - 1 \right), \\ v^2 &= \frac{\omega^2}{\xi^2}, \quad v_T^2 = \frac{c_{44}}{\rho^0}. \end{aligned} \tag{5.121}$$

The substitution of Eq. (5.120) into Eq. (5.119) leads to

$$c_{44} U_3 \eta \cos \eta h = 0. \tag{5.122}$$

For nontrivial solutions we must have

$$\cos \eta h = 0, \tag{5.123}$$

which determines the dispersion relations of the waves as

$$\eta h = \frac{n\pi}{2}, \quad n = 1, 3, 5 \ldots, \tag{5.124}$$

or

$$\eta^2 h^2 = \frac{\rho^0 h^2 \omega^2}{c_{44}} - \xi^2 h^2 = \frac{n^2 \pi^2}{4}, \tag{5.125}$$

where Eq. (5.121) has been used. Equation (5.125) can be written into the following dimensionless form:

$$\Omega^2 - X^2 = n^2, \tag{5.126}$$

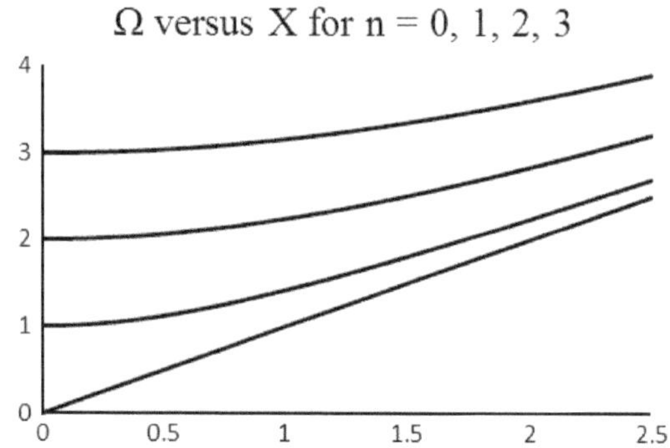

Fig. 5.12. Dispersion curves from Eq. (5.126) with $n = 0, 1, 2$ and 3.

where the dimensionless frequency Ω and the dimensionless wave number X in the x_1 direction are defined by

$$\Omega = \omega \Big/ \left(\frac{\pi}{2h}\sqrt{\frac{c_{44}}{\rho^0}}\right), \quad X = \xi \Big/ \left(\frac{\pi}{2h}\right). \tag{5.127}$$

For symmetric waves, we consider

$$u_3 = U_3 \cos \eta x_2 \cos(\xi x_1 - \omega t). \tag{5.128}$$

Through a similar analysis, Eq. (5.128) also leads to Eq. (5.126) with n assuming even integers. The dispersion curves described by Eq. (5.126) are shown in Fig. 5.12.

5.12 Love Wave

Consider antiplane motions with $u_1 = 0$, $u_2 = 0$ and $u_3 = u_3(x_1, x_2, t)$ in an elastic plate on an elastic half-space of another material as shown in Fig. 5.13. The governing equations for the half-space and the plate are, respectively,

$$\begin{aligned} c_{44}\nabla^2 u_3 &= \rho^0 \ddot{u}_3, \quad x_2 > 0, \\ \hat{c}_{44}\nabla^2 u_3 &= \hat{\rho}^0 \ddot{u}_3, \quad -h < x_2 < 0, \end{aligned} \tag{5.129}$$

where $\hat{\rho}^0$ and $\hat{c}_{44}$ are the mass density and shear elastic constant of the plate. We look for solutions satisfying

$$u_3 \to 0, \quad x_2 \to +\infty. \tag{5.130}$$

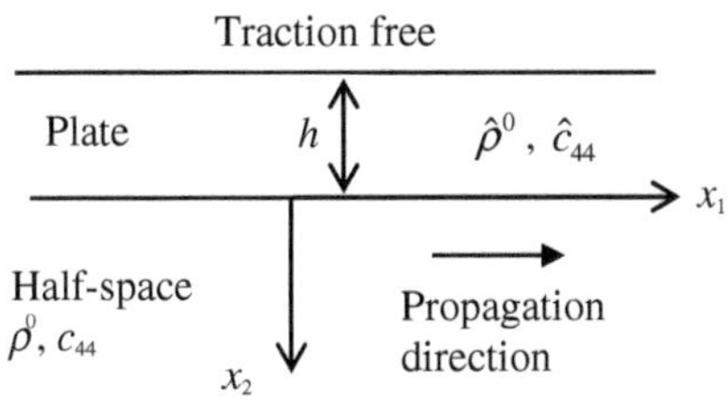

Fig. 5.13. A plate on a half-space of a different material.

For the half-space in $x_2>0$, the solution to Eq. $(5.129)_1$ satisfying Eq. (5.130) can be written as

$$u_3 = A\exp(-\eta x_2)\cos(\xi x_1 - \omega t). \tag{5.131}$$

where, for Eq. (5.131) to satisfy Eq. $(5.129)_1$, we have the following relationship among the parameters of Eq. (5.131):

$$\eta^2 = \xi^2 - \frac{\rho^0\omega^2}{c_{44}} = \xi^2\left(1 - \frac{v^2}{v_T^2}\right) > 0. \tag{5.132}$$

In Eq. (5.132), we have denoted

$$v^2 = \frac{\omega^2}{\xi^2}, \quad v_T^2 = \frac{c_{44}}{\rho^0}. \tag{5.133}$$

The stress component needed for the interface continuity condition is given by

$$\sigma_{23} = c_{44}u_{3,2} = -c_{44}A\eta\exp(-\eta x_2)\cos(\xi x_1 - \omega t). \tag{5.134}$$

For the plate in $-h<x_2<0$, let

$$u_3 = (\hat{A}\cos\hat{\eta}x_2 + \hat{B}\sin\hat{\eta}x_2)\cos(\xi x_1 - \omega t), \tag{5.135}$$

where, from Eq. $(5.129)_2$,

$$\hat{\eta}^2 = \frac{\hat{\rho}^0\omega^2}{\hat{c}} - \xi^2 = \xi^2\left(\frac{v^2}{\hat{v}_T^2} - 1\right), \tag{5.136}$$

and

$$\hat{v}_T^2 = \frac{\hat{c}}{\hat{\rho}^0}. \tag{5.137}$$

For the boundary and continuity conditions of the plate, we need,

$$\begin{aligned}\sigma_{23} &= \hat{c}_{44} u_{3,2} \\ &= \hat{c}_{44}(-\hat{A}\hat{\eta}\sin\hat{\eta}x_2 + \hat{B}\hat{\eta}\cos\hat{\eta}x_2)\cos(\xi x_1 - \omega t). \end{aligned} \tag{5.138}$$

The continuity and boundary conditions are, after the cancellation of a common factor of $\cos(\xi x_1 - \omega t)$,

$$\begin{aligned} u_3(0^+) &= A = \hat{A} = u_3(0^-), \\ \sigma_{23}(0^+) &= -c_{44}A\eta = \hat{c}_{44}\hat{B}\hat{\eta} = \sigma_{23}(0^-), \\ \sigma_{23}(-h) &= \hat{c}_{44}(\hat{A}\hat{\eta}\sin\hat{\eta}h + \hat{B}\hat{\eta}\cos\hat{\eta}h) = 0. \end{aligned} \tag{5.139}$$

Using Eqs. $(5.139)_1$ to eliminate A, we obtain

$$\begin{aligned} -c_{44}\hat{A}\eta - \hat{c}_{44}\hat{B}\hat{\eta} = 0, \\ \hat{A}\hat{\eta}\sin\hat{\eta}h + \hat{B}\hat{\eta}\cos\hat{\eta}h = 0. \end{aligned} \tag{5.140}$$

For nontrivial solutions, the determinant of the coefficient matrix of Eq. (5.140) must vanish, i.e.,

$$\begin{vmatrix} -c_{44}\eta & -\hat{c}_{44}\hat{\eta} \\ \hat{\eta}\sin\hat{\eta}h & \hat{\eta}\cos\hat{\eta}h \end{vmatrix} = 0, \tag{5.141}$$

or

$$\frac{\eta}{\xi} - \frac{\hat{c}_{44}}{c_{44}}\frac{\hat{\eta}}{\xi}\tan\hat{\eta}h = 0. \tag{5.142}$$

Substituting from Eqs. (5.132) and (5.136), we obtain

$$\sqrt{1 - \frac{v^2}{v_T^2}} - \frac{\hat{c}_{44}}{c_{44}}\sqrt{\frac{v^2}{\hat{v}_T^2} - 1}\tan\left[\xi h\sqrt{\frac{v^2}{\hat{v}_T^2} - 1}\right] = 0, \tag{5.143}$$

which determines the Love wave speed v as a function of the wave number ξ. The waves determined by Eq. (5.143) are dispersive, i.e., v depends on ξ. The dispersion relations for Love waves are real and multi-valued when $\hat{v}_T^2 < v^2 < v_T^2$, i.e., when the material constants are such that the shear wave speed of the plate is smaller than that of the half-space.

5.13 Vibration of a Rectangular Body

Consider antiplane free vibrations of the rectangular body [17] in Fig. 5.14. The boundary is fixed. $u_1 = 0$, $u_2 = 0$ and $u_3 = u(x_1,x_2,t)$. The governing equation and boundary conditions are

$$\begin{aligned} &\mu(u_{,11} + u_{,22}) = \rho^0 \ddot{u} \quad \text{in} \quad A, \\ &u = 0, \quad x_1 = 0, a \quad \text{or} \quad x_2 = 0, b. \end{aligned} \tag{5.144}$$

We look for solutions in the following form:

$$u(x_1, x_2, t) = u(x_1, x_2) \exp(i\omega t). \tag{5.145}$$

Substituting Eq. (5.145) into Eq. $(5.144)_1$, we have

$$u_{,11} + u_{,22} + \beta^2 u = 0, \tag{5.146}$$

where

$$\beta^2 = \frac{\rho^0}{\mu}\omega^2. \tag{5.147}$$

For solutions of Eq. (5.146), we use the well-known method of separation of variables in partial differential equations. Let

$$u(x, y) = X(x)Y(y), \tag{5.148}$$

which, when substituted into Eq. (5.146), leads to

$$\frac{1}{X}\frac{d^2X}{dx^2} + \frac{1}{Y}\frac{d^2Y}{dy^2} + \beta^2 = 0. \tag{5.149}$$

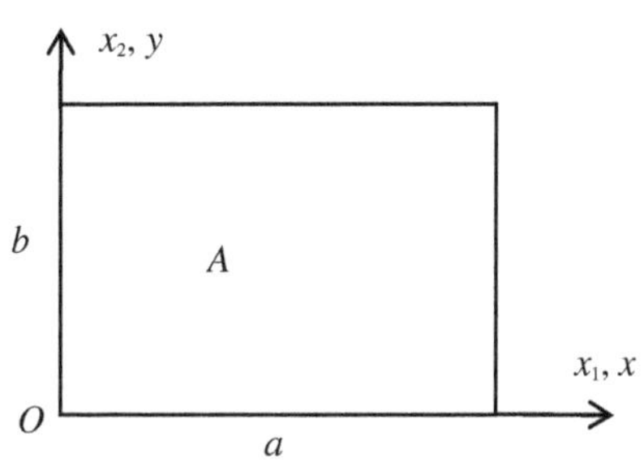

Fig. 5.14. A rectangular body and coordinate system.

Equation (5.149) implies that

$$\frac{1}{X}\frac{d^2X}{dx^2} = -\alpha^2, \quad \frac{1}{Y}\frac{d^2Y}{dy^2} = -\gamma^2, \tag{5.150}$$

with

$$\alpha^2 + \gamma^2 = \beta^2. \tag{5.151}$$

From Eq. (5.150), we obtain

$$\begin{aligned} X &= C_1 \sin \alpha x + C_2 \cos \alpha x, \\ Y &= C_3 \sin \gamma y + C_4 \cos \gamma y. \end{aligned} \tag{5.152}$$

Then, from Eq. (5.148),

$$\begin{aligned} u = {} & A_1 \sin \alpha x \sin \gamma y + A_2 \sin \alpha x \cos \gamma y \\ & + A_3 \cos \alpha x \sin \gamma y + A_4 \cos \alpha x \cos \gamma y. \end{aligned} \tag{5.153}$$

For $u = 0$ at $x = 0$, we must have

$$A_3 \sin \gamma y + A_4 \cos \gamma y = 0. \tag{5.154}$$

The case of $\gamma = 0$ implies that $Y = 0$. When $\gamma \neq 0$, Eq. (5.154) leads to

$$A_3 = A_4 = 0. \tag{5.155}$$

Hence,

$$u = A_1 \sin \alpha x \sin \gamma y + A_2 \sin \alpha x \cos \gamma y. \tag{5.156}$$

For $u = 0$ at $x = a$, we must have

$$A_1 \sin \alpha a \sin \gamma y + A_2 \sin \alpha a \cos \gamma y = 0. \tag{5.157}$$

For nontrivial solutions,

$$\sin \alpha a = 0, \tag{5.158}$$

or

$$\alpha a = m\pi, \quad m = 1, 2, \ldots. \tag{5.159}$$

For $u = 0$ at $y = 0$, Eq. (5.156) leads to

$$A_2 \sin \alpha x = 0, \tag{5.160}$$

which implies that

$$A_2 = 0. \tag{5.161}$$

Hence,

$$u = A_1 \sin \alpha x \sin \gamma y. \tag{5.162}$$

For $u = 0$ at $y = b$, we must have

$$A_1 \sin \alpha x \sin \gamma b = 0, \tag{5.163}$$

which leads to

$$\begin{aligned} &\sin \gamma b = 0, \\ &\gamma b = n\pi, \quad n = 1, 2, \ldots. \end{aligned} \tag{5.164}$$

Hence, the resonance frequencies and modes are

$$u = A_1 \sin \frac{m\pi x}{a} \sin \frac{n\pi y}{b}, \tag{5.165}$$

$$\beta^2 = \alpha^2 + \gamma^2 = \left(\frac{m\pi}{a}\right)^2 + \left(\frac{n\pi}{b}\right)^2, \tag{5.166}$$

$$\omega^2 = \frac{\mu}{\rho^0}\beta^2 = \frac{\mu}{\rho^0}\left[\left(\frac{m\pi}{a}\right)^2 + \left(\frac{n\pi}{b}\right)^2\right]. \tag{5.167}$$

5.14 Spherical Waves

Consider spherical waves in an unbounded domain [18]. In spherical coordinates (see Appendix 3), let

$$u_r = u(r, t), \quad u_\theta = u_\phi = 0. \tag{5.168}$$

From the equations in Appendix 3, we have the following strain field:

$$\begin{aligned} &\varepsilon_{rr} = \frac{du}{dr}, \quad \varepsilon_{\theta\theta} = \varepsilon_{\phi\phi} = \frac{u}{r}, \\ &\varepsilon_{r\theta} = \varepsilon_{\theta\phi} = \varepsilon_{\phi r} = 0, \end{aligned} \tag{5.169}$$

stress field:

$$\sigma_{rr} = (\lambda + 2\mu)\frac{du}{dr} + \lambda 2\frac{u}{r},$$
$$\sigma_{\theta\theta} = \sigma_{\phi\phi} = (\lambda + 2\mu)\frac{u}{r} + \lambda\left(\frac{du}{dr} + \frac{u}{r}\right), \tag{5.170}$$
$$\sigma_{r\theta} = \sigma_{\theta\phi} = \sigma_{\phi r} = 0,$$

and equation of motion:

$$\frac{d\sigma_{rr}}{dr} + \frac{1}{r}(2\sigma_{rr} - \sigma_{\theta\theta} - \sigma_{\phi\phi}) = \rho^0 \ddot{u}. \tag{5.171}$$

With substitutions from Eq. (5.170), the equation of motion takes the following form:

$$\frac{d^2u}{dr^2} + \frac{2}{r}\frac{du}{dr} - \frac{2}{r^2}u = \frac{1}{c_1^2}\ddot{u}, \tag{5.172}$$

where

$$c_1^2 = \frac{\lambda + 2\mu}{\rho^0}. \tag{5.173}$$

Let,

$$u = \frac{\partial \phi}{\partial r}. \tag{5.174}$$

Equation (5.172) becomes

$$\frac{\partial}{\partial r}\left\{\frac{1}{r}\left[\frac{\partial^2(r\phi)}{\partial r^2} - \frac{1}{c_1^2}\frac{\partial^2(r\phi)}{\partial t^2}\right]\right\} = 0, \tag{5.175}$$

which is satisfied if

$$\frac{\partial^2(r\phi)}{\partial r^2} - \frac{1}{c_1^2}\frac{\partial^2(r\phi)}{\partial t^2} = 0. \tag{5.176}$$

Equation (5.176) admits the d'Alembert solution:

$$\phi = \frac{1}{r}[f(r - c_1 t) + g(r + c_1 t)]. \tag{5.177}$$

5.15 Equations for Nonlinear Thermoviscoelasticity

The field equations of nonlinear thermoviscoelasticity are from Eqs. (5.5), (5.4) and (4.70):

$$\begin{aligned} &\rho^0 = \rho J, \\ &K_{Lk,L} + \rho^0 f_k = \rho^0 \ddot{y}_k, \\ &\rho^0 \theta \dot{\eta} = P^D_{KL} \dot{E}_{KL} + \rho^0 r - Q_{K,K}. \end{aligned} \tag{5.178}$$

The constitutive relations are from Eqs. (4.65), (4.64), (4.68), (4.67), (4.62), (4.69) and (4.21):

$$\chi = \chi(E_{KL}; \theta), \tag{5.179}$$

$$\eta = -\frac{\partial \chi}{\partial \theta}, \tag{5.180}$$

$$P^R_{KL} = \rho^0 \frac{\partial \chi}{\partial E_{KL}}, \tag{5.181}$$

$$P^D_{KL} = P^D_{KL}(E_{KL}; \theta; \Theta_K; \dot{E}_{KL}), \tag{5.182}$$

$$Q_K = Q_K(E_{KL}; \theta; \Theta_K; \dot{E}_{KL}), \tag{5.183}$$

$$P^D_{KL} \dot{E}_{KL} - \frac{1}{\theta} Q_K \Theta_K \geq 0, \tag{5.184}$$

$$P_{KL} = P^R_{KL} + P^D_{KL}, \tag{5.185}$$

$$K_{Lj} = y_{j,M} P_{ML}. \tag{5.186}$$

With successive substitutions, Eq. $(5.178)_{2,3}$ may be written as four equations for $y_i(\mathbf{X}, t)$ and $\theta(\mathbf{X}, t)$. The equation for the conservation of mass in Eq. $(5.178)_1$ can be used to determine ρ separately from the other equations. On the boundary surface S, the thermal boundary condition may be either a prescribed temperature or a heat flux

$$N_L Q_L = \bar{Q}. \tag{5.187}$$

5.16 Equations for Linear Thermoviscoelasticity

For linear thermoviscoelasticity, we have the conservation of mass from Eq. (5.9):

$$\rho^0 = \rho(1 + u_{k,k}). \tag{5.188}$$

Linear momentum equation from Eq. (5.10):

$$\sigma_{ij,i} + b_j = \rho^0 \ddot{u}_j. \tag{5.189}$$

Dissipation equation from Eq. (4.81):

$$\rho^0 T_0 \dot{\eta} = \sigma_{ij}^D \dot{\varepsilon}_{ij} + \rho^0 r - q_{i,i}. \tag{5.190}$$

Strain-displacement relation from Eq. (5.12):

$$\varepsilon_{ij} = (u_{i,j} + u_{j,i})/2. \tag{5.191}$$

Free energy from Eq. (4.75):

$$\rho^0 \chi = \frac{1}{2} c_{ijkl} \varepsilon_{ij} \varepsilon_{kl} - \frac{1}{2} \frac{\rho^0 c}{T_0} T^2 - \beta_{kl} \varepsilon_{kl} T. \tag{5.192}$$

Constitutive relations from Eqs. (4.76), (4.78) and (4.79):

$$\sigma_{ij}^R = \rho^0 \frac{\partial \chi}{\partial \varepsilon_{ij}} = c_{ijkl} \varepsilon_{kl} - \beta_{ij} T,$$

$$\rho^0 \eta = -\rho^0 \frac{\partial \chi}{\partial \theta} = \frac{\rho^0 c}{T_0} T + \beta_{kl} \varepsilon_{kl}, \tag{5.193}$$

$$\sigma_{ij}^D = \mu_{ijkl} \dot{\varepsilon}_{kl}, \tag{5.194}$$

$$q_k = -\kappa_{kl} T_{,l}, \tag{5.195}$$

and

$$\sigma_{ij} = \sigma_{ij}^R + \sigma_{ij}^D. \tag{5.196}$$

Equations (5.194) and (5.195) are restricted by Eq. (4.80):

$$\sigma_{ij}^D \dot{\varepsilon}_{ij} - \frac{q_i T_{,i}}{T_0} \geq 0. \tag{5.197}$$

Problems

5.1. For linear elastostatics, show Clapeyron's theorem:

$$\frac{1}{2} \int_V b_i u_i dV + \frac{1}{2} \int_S t_i u_i dS = \int_V U dV.$$

5.2. Consider the following two linear elastostatic problems over the same region:

$$\begin{aligned} \sigma_{ji,j}^{(1)} + b_i^{(1)} &= 0 \quad \text{in} \quad V, \\ \sigma_{ij}^{(1)} &= c_{ijkl} u_{k,l}^{(1)} \quad \text{in} \quad V, \\ n_j \sigma_{ji}^{(1)} &= t_i^{(1)} \quad \text{on} \quad S, \end{aligned}$$

and

$$\begin{aligned} \sigma_{ji,j}^{(2)} + b_i^{(2)} &= 0 \quad \text{in} \quad V, \\ \sigma_{ij}^{(2)} &= c_{ijkl} u_{k,l}^{(2)} \quad \text{in} \quad V, \\ n_j \sigma_{ji}^{(2)} &= t_i^{(2)} \quad \text{on} \quad S. \end{aligned}$$

Show Betti's reciprocal theorem, i.e.,

$$\int_V b_i^{(1)} u_i^{(2)} dV + \int_S t_i^{(1)} u_i^{(2)} dS = \int_V b_i^{(2)} u_i^{(1)} dV + \int_S t_i^{(2)} u_i^{(1)} dS.$$

5.3. For linear elastostatics, obtain the Beltrami–Michell equation:

$$\nabla^2 \sigma_{ij} + \frac{1}{1+\nu} \sigma_{kk,ij} + \frac{\nu}{1-\nu} \delta_{ij} b_{k,k} + b_{i,j} + b_{j,i} = 0.$$

5.4. For linear elastostatics, show that the stationary condition of the variational functional

$$\Pi(\mathbf{u}) = \int_V [-U(\varepsilon) + b_i u_i] dV + \int_{S_\sigma} \bar{t}_i u_i dS,$$

under the following constraints

$$\begin{aligned} \varepsilon_{ij} &= (u_{i,j} + u_{j,i})/2, \\ u_i &= \bar{u}_i \quad \text{on} \quad S_u, \end{aligned}$$

leads to

$$\begin{aligned} \sigma_{ji,j} + b_i &= 0 \quad \text{in } V, \\ n_j \sigma_{ji} &= \bar{t}_i \quad \text{on} \quad S_\sigma. \end{aligned}$$

5.5. For linear elastostatics, determine the displacement field corresponding to zero strains.

5.6. A cylindrical and elliptic hole in an infinite body is filled with a uniform internal pressure p. Determine the traction boundary conditions on the hole surface. $u_3 = 0$ and $\partial/\partial x_3 = 0$.

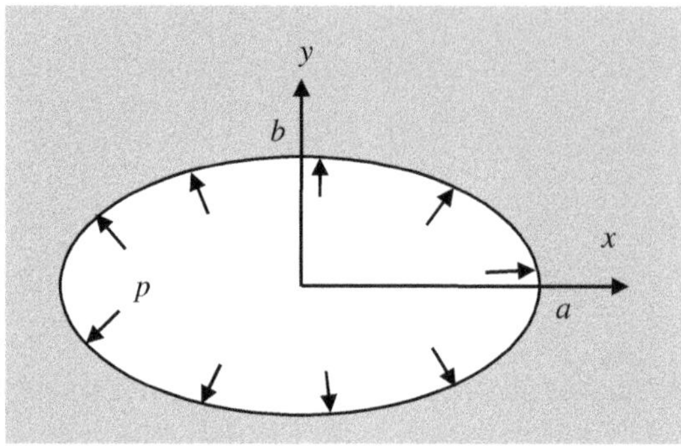

5.7. Consider the rectangular body as shown with $u_3 = 0$ and $\partial/\partial x_3 = 0$. The body is fixed at $x_2 = 0$. What are the boundary conditions for the displacement formulation? If, to make use of the symmetry and/or antisymmetry in the problem, the right half of the body ($x_1 > 0$) is taken to be analyzed, what are the boundary conditions at the new boundary where $x_1 = 0$?

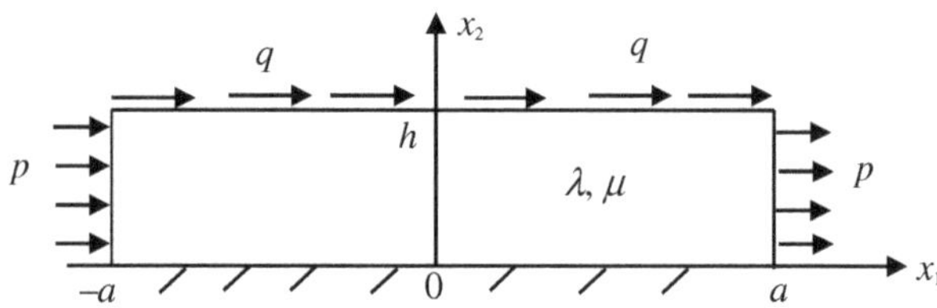

5.8. Consider the following inclusion problem with $u_3 = 0$ and $\partial/\partial x_3 = 0$. A uniaxial tensile stress $p =$ constant is applied at $x = \pm\infty$. Specify the governing equations, boundary conditions and continuity conditions needed for analyzing this problem.

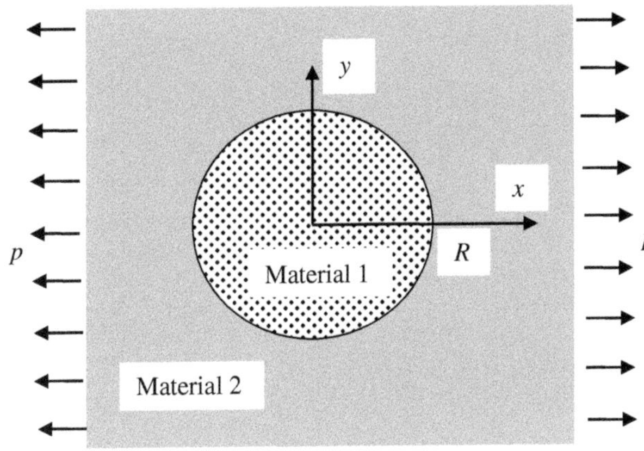

5.9. Obtain the solution to the problem of an elastic layer under normal surface traction and its own weight (see Sections 5.3 and 5.4).

5.10. Consider an elastic layer with a uniform thickness h. The layer is unbounded in the (x_1, x_2) plane. It is fixed at $x_3 = 0$. At $x_3 = h$, the traction vector is given to be $\bar{t}_1 = q$, $\bar{t}_2 = 0$, and $\bar{t}_3 = p$. Determine the displacement and stress fields in the layer.

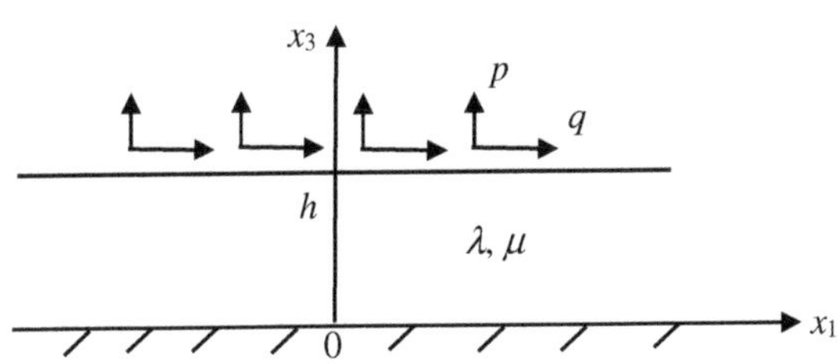

5.11. Consider two perfectly bonded layers of different materials with thicknesses h_1 and h_2. The bottom layer is fixed at $x_3 = 0$. At $x_3 = h_1 + h_2$, the displacement vector is given to be $u_1 = 0$, $u_2 = 0$, and $u_3 = \delta$, where δ is a constant. Determine the displacement field in the layers.

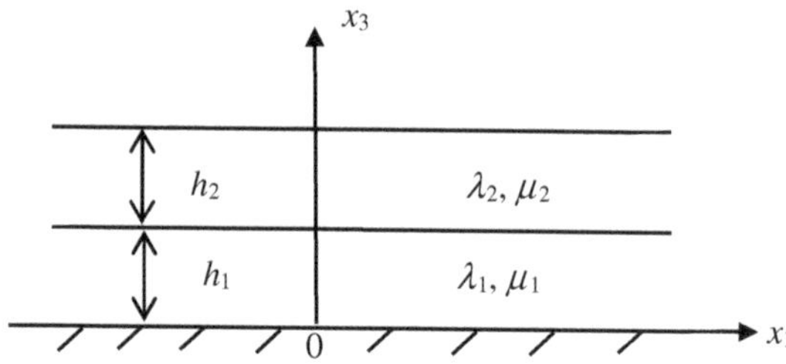

5.12. Study an infinitely long circular cylindrical tube under uniform inner and outer pressures.

5.13. Study a spherical shell under uniform inner and outer pressures.

5.14. Consider the circular shaft and its cross section shown. The radius of the cross section is R. It is in torsion under a torque T only. Obtain the solution using

$$u_1 = -\theta x_3 x_2, \quad u_2 = \theta x_3 x_1, \quad u_3 = 0,$$

where θ is an undetermined constant (angle of twist per unit length).

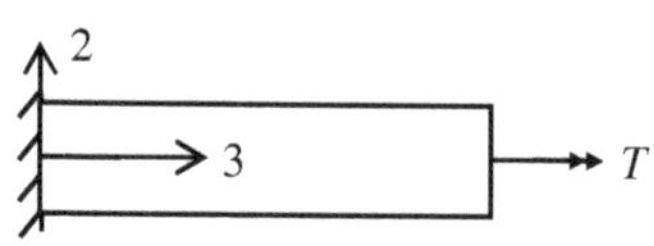

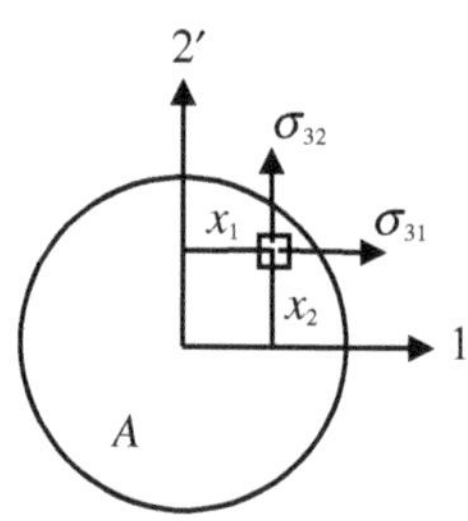

5.15. Show that for an isotropic elastic body without body force, Navier's equation is satisfied by $u_k = \varphi_{,k} + \varepsilon_{kij}\psi_{j,i}$ provided that

$$c_1^2\varphi_{,kk} = \ddot{\varphi}, \quad c_2^2\psi_{j,kk} = \ddot{\psi}_j,$$
$$c_1^2 = \frac{\lambda + 2\mu}{\rho^0}, \quad c_2^2 = \frac{\mu}{\rho^0}.$$

5.16. For plane waves in an isotropic and unbounded domain, show that the longitudinal wave speed c_1 is larger than the transverse wave speed c_2 (see Section 5.7).
5.17. Obtain dispersion relations for antiplane waves in a plate with fixed surfaces.
5.18. Study antiplane free vibrations of a circular body with radius R.

Chapter 6

Fluids

In this chapter, we collect the relevant equations for Stokesian fluids and then reduce them to the equations for Newtonian fluids, ideal fluids, and the equations for acoustics. These are followed by the solutions of a series of problems of fluid statics and dynamics.

6.1 Equations for Stokesian Fluids

The field equations in spatial forms are taken from Eqs. (3.128), (3.129) and (4.104):

$$\dot{\rho} + \rho v_{i,i} = 0, \tag{6.1}$$

$$\rho \dot{v}_k = \tau_{ik,i} + \rho f_k, \tag{6.2}$$

$$\rho\theta \frac{d\eta}{dt} = \tau^D_{kl} d_{kl} + \rho r - q_{j,j}. \tag{6.3}$$

The free energy χ, the definition of the thermodynamic pressure p, the decomposition of the stress tensor $\boldsymbol{\tau}$ and the constitutive relations below are taken from Eqs. (4.95)–(4.103):

$$\chi = \chi(\rho^{-1}; \theta), \tag{6.4}$$

$$p = -\frac{\partial \chi}{\partial(\rho^{-1})}, \tag{6.5}$$

$$\eta = -\frac{\partial \chi}{\partial \theta}, \tag{6.6}$$

$$\tau_{kl} = -p\delta_{kl} + \tau_{kl}^D, \tag{6.7}$$

$$\tau_{kl}^D = \tau_{kl}^D(\rho^{-1}; \mathbf{d}; \nabla\theta; \theta), \tag{6.8}$$

$$q_k = q_k(\rho^{-1}; \mathbf{d}; \nabla\theta; \theta), \tag{6.9}$$

$$\tau_{kl}^D d_{kl} - \frac{q_i \theta_{,i}}{\theta} \geq 0. \tag{6.10}$$

With successive substitutions, we can write Eqs. (6.1)–(6.3) as five equations for $\mathbf{v}$, θ and ρ.

6.2 Equations for Newtonian Fluids

For Newtonian fluids, the governing equations are as follows. The only difference from the previous section is that the viscous stress tensor $\boldsymbol{\tau}^D$ and the heat flux vector $\mathbf{q}$ are linear in the velocity gradient and temperature gradient, respectively.

$$\dot{\rho} + \rho v_{i,i} = 0, \tag{6.11}$$

$$\rho \dot{v}_k = \tau_{ik,i} + \rho f_k, \tag{6.12}$$

$$\rho\theta \frac{d\eta}{dt} = \tau_{kl}^D d_{kl} + \rho r - q_{j,j}, \tag{6.13}$$

$$\chi = \chi(\rho^{-1}; \theta), \tag{6.14}$$

$$p = -\frac{\partial \chi}{\partial(\rho^{-1})}, \tag{6.15}$$

$$\eta = -\frac{\partial \chi}{\partial \theta}, \tag{6.16}$$

$$\tau_{kl} = -p\delta_{kl} + \tau_{kl}^D, \tag{6.17}$$

$$\tau_{ij}^D = \lambda v_{k,k}\delta_{ij} + \mu(v_{j,i} + v_{i,j}), \tag{6.18}$$

$$q_k = -\kappa_{kl}\theta_{,l}. \tag{6.19}$$

In general, λ and μ may still depend on ρ and θ. When λ and μ are constants, Eq. (6.12) can be written as

$$(\lambda + \mu)v_{k,kl} + \mu v_{l,kk} - p_{,l} + \rho f_l = \rho \dot{v}_l, \tag{6.20}$$

which is known as Navier–Stokes equation. Under Stokes' assumption that $\lambda = -2\mu/3$ (see Eq. (4.110)), Eq. (6.20) becomes

$$\frac{\mu}{3} v_{k,kl} + \mu u_{l,kk} - p_{,l} + \rho f_l = \rho \dot{v}_l. \tag{6.21}$$

6.3 Equations for Ideal Fluids

For ideal fluids, the field equations are

$$\dot{\rho} + \rho v_{i,i} = 0, \tag{6.22}$$

$$\rho \dot{v}_k = \tau_{ik,i} + \rho f_k. \tag{6.23}$$

The internal energy density and pressure are taken from Eqs. (4.117) and (4.119):

$$\varepsilon = \varepsilon(\rho^{-1}), \tag{6.24}$$

$$p = -\frac{\partial \varepsilon}{\partial(\rho^{-1})}. \tag{6.25}$$

The stress is given by Eq. (4.122):

$$\tau_{ij} = -p\delta_{ij}. \tag{6.26}$$

Equation (6.23) can be written as

$$\frac{\partial \mathbf{v}}{\partial t} + (\mathbf{v} \cdot \nabla)\mathbf{v} = -\frac{1}{\rho}\nabla p + \mathbf{f}, \tag{6.27}$$

which is known as Euler's equation.

When the fluid is barotropic with [19]

$$\rho = \rho(p), \tag{6.28}$$

we introduce P such that

$$P(p) = \int \frac{dp}{\rho}. \tag{6.29}$$

Then

$$\nabla P = \frac{\nabla p}{\rho}. \tag{6.30}$$

Consider the case when the body force has a potential, i.e.,

$$\mathbf{f} = -\nabla V. \tag{6.31}$$

Then Euler's equation becomes

$$\frac{d\mathbf{v}}{dt} = -\nabla P - \nabla V. \tag{6.32}$$

Dotting both sides by $\mathbf{v}$ gives

$$\mathbf{v} \cdot \frac{d\mathbf{v}}{dt} = -\mathbf{v} \cdot \nabla(P + V), \tag{6.33}$$

$$\frac{d}{dt}\left(\frac{1}{2}\mathbf{v} \cdot \mathbf{v}\right) = -\mathbf{v} \cdot \nabla(P + V). \tag{6.34}$$

For steady flows with $\partial/\partial t = 0$, we have

$$\frac{d(\ )}{dt} = \mathbf{v} \cdot \nabla(\ ). \tag{6.35}$$

Equation (6.34) becomes

$$\frac{d}{dt}\left(\frac{1}{2}\mathbf{v} \cdot \mathbf{v} + P + V\right) = 0. \tag{6.36}$$

Hence, following a fluid particle, we have

$$\frac{1}{2}\mathbf{v} \cdot \mathbf{v} + P + V = C, \tag{6.37}$$

where C is a constant. For steady flows, the path lines and stream lines coincide with each other. Then Eq. (6.37) is true following a stream line. Equation (6.37) is the well known Bernoulli's Equation.

6.4 Equations for Acoustics

Consider an ideal and barotropic fluid with

$$p = p(\rho). \tag{6.38}$$

The conservation of mass and the linear momentum equation are

$$\dot{\rho} + \rho v_{i,i} = 0, \tag{6.39}$$

$$\rho \dot{v}_k = -p_{,k}, \tag{6.40}$$

where the body force is assumed to be absent. Let a static and uniform reference state be

$$\mathbf{v} = 0, \quad p = p_0, \quad \rho = \rho_0, \quad p_0 = p(\rho_0). \tag{6.41}$$

Under small disturbances, the fields become

$$\mathbf{v} = \tilde{\mathbf{v}}, \quad p = p_0 + \tilde{p}, \quad \rho = \rho_0 + \tilde{\rho}. \tag{6.42}$$

From Eqs. (6.38) and (6.41),

$$p_0 + \tilde{p} = p(\rho_0 + \tilde{\rho}) \quad \Rightarrow \quad \tilde{p} = \left.\frac{\partial p}{\partial \rho}\right|_{\rho_0} \tilde{\rho} = c_0^2 \tilde{\rho}, \tag{6.43}$$

where we have denoted

$$c_0^2 = \left.\frac{\partial p}{\partial \rho}\right|_{\rho_0}. \tag{6.44}$$

At the same time, Eqs. (6.39) and (6.40) reduce to

$$\frac{\partial \tilde{\rho}}{\partial t} + \rho_0 \tilde{v}_{i,i} = 0, \tag{6.45}$$

$$\rho_0 \frac{\partial \tilde{v}_k}{\partial t} = -\tilde{p}_{,k}. \tag{6.46}$$

Eliminating $\tilde{\mathbf{v}}$ and $\tilde{p}$ from Eqs. (6.43), (6.45) and (6.46), we obtain a single wave equation for $\tilde{\rho}$

$$\frac{\partial^2 \tilde{\rho}}{\partial t^2} = c_0^2 \tilde{\rho}_{,kk}, \tag{6.47}$$

which shows that c_0 is the acoustic wave speed. With the use of Eq. (6.43), we can also write Eq. (6.47) as a wave equation for $\tilde{p}$:

$$\frac{\partial^2 \tilde{p}}{\partial t^2} = c_0^2 \tilde{p}_{,kk}. \tag{6.48}$$

6.5 Hydrostatic Pressure

Consider the equilibrium of a semi-infinite fluid under gravity (see Fig. 6.1). The fluid is incompressible with a constant ρ. The body force is given by

$$f_x = f_y = 0, \quad f_z = -g. \tag{6.49}$$

Euler's equation reduces to

$$\nabla p = \rho \mathbf{f} = -\rho g \mathbf{k}, \tag{6.50}$$

or

$$\nabla (p + \rho g z) = 0. \tag{6.51}$$

Hence,

$$p + \rho g z = C, \tag{6.52}$$

where C is an undetermined constant. Assuming that the atmospheric pressure at the fluid surface where $z = 0$ is p_0, then

$$p|_{z=0} = p_0 \quad \Rightarrow \quad C = p_0. \tag{6.53}$$

Hence,

$$p = p_0 - \rho g z. \tag{6.54}$$

In the special case when $g = 0$, we have $p = p_0$ which is an example of Pascal's law stating that "a pressure change at any point in a confined incompressible fluid is transmitted throughout the fluid such that the same change occurs everywhere".

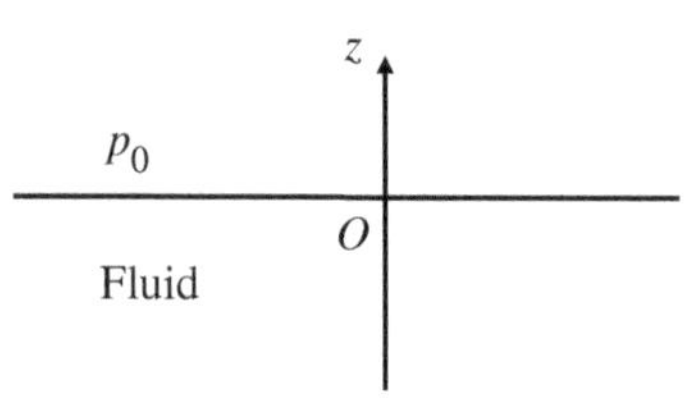

Fig. 6.1. A semi-infinite fluid under gravity and surface pressure.

6.6 Archimedes' Principle

Consider a rigid body fully submerged in an incompressible fluid under gravity as shown in Fig. 6.2. Archimedes' principle states that "a body at rest in a fluid is acted upon by a force pushing upward called the buoyant force, which is equal to the weight of the fluid that the body displaces. If the body is completely submerged, the volume of fluid displaced is equal to the volume of the body". We prove the principle below. From Eq. (6.54), the pressure field is given by

$$p = p_0 - \rho g z = p_0 - \rho g x_3. \tag{6.55}$$

The resultant of the pressure on the surface of the body can be calculated from

$$\begin{aligned} F_i &= \int_S -p n_i dS = \int_S -(p_0 - \rho g x_3) n_i dS \\ &= \int_V -(p_0 - \rho g x_3)_{,i} dV = \int_V \rho g \delta_{3i} dV = \rho g V \delta_{3i}, \end{aligned} \tag{6.56}$$

or

$$F_1 = 0, \quad F_2 = 0, \quad F_3 = \rho g V, \tag{6.57}$$

which proves the principle. The case of a partially submerged body is left as an exercise.

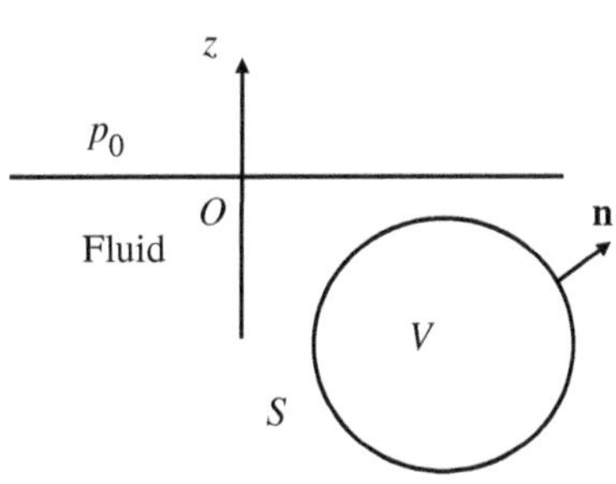

Fig. 6.2. A completely submerged body under hydrostatic pressure.

6.7 A Fluid in an Accelerating Container

Consider a container accelerating at a constant acceleration a (see Fig. 6.3). It is filled with an incompressible fluid which is in equilibrium with respect to the container. The coordinate system is fixed to the container and hence is non-inertial. In the accelerating coordinate system, D'Alembert's principle may be used which states that the sum of the applied forces and the inertial forces is zero. The inertial force is the negative of the mass times the acceleration. Therefore, the effective body forces are

$$f_x = -a, \quad f_y = 0, \quad f_z = -g. \tag{6.58}$$

Then Euler's equation reduces to

$$\nabla p = \rho \mathbf{f} = -\rho a \mathbf{i} - \rho g \mathbf{k}, \tag{6.59}$$

which can be written as

$$\nabla(p + \rho a x + \rho g z) = 0. \tag{6.60}$$

Therefore,

$$p + \rho a x + \rho g z = C, \tag{6.61}$$

where C is an undetermined constant. At the surface of the fluid $p = p_0$, i.e.,

$$p_0 + \rho a x + \rho g z = C, \tag{6.62}$$

which is the equation of the fluid surface. It is a plane. If when $a = 0$ the fluid surface is at $z = 0$, then $C = p_0$.

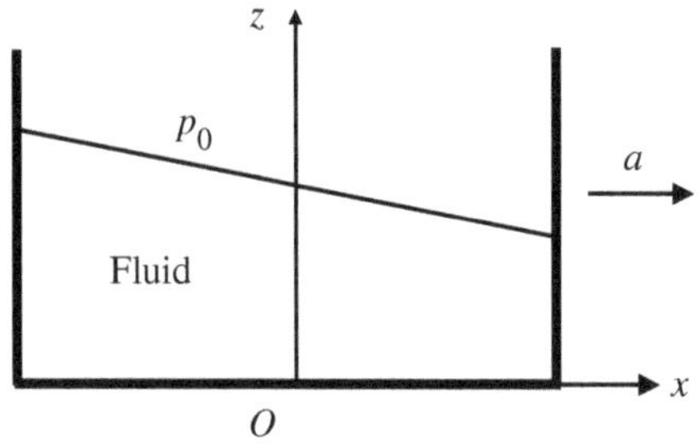

Fig. 6.3. A fluid in an accelerating container.

6.8 A Fluid in a Rotating Container

Consider an incompressible fluid in a container rotating at a constant angular rate ω as shown in Fig. 6.4. According to D'Alembert's principle, in the co-rotating reference frame, the effective body forces are

$$f_x = \omega^2 x, \quad f_y = \omega^2 y, \quad f_z = -g. \tag{6.63}$$

Euler's equation takes the following form:

$$\nabla p = \rho \mathbf{f} = \rho\omega^2 x\mathbf{i} + \rho\omega^2 y\mathbf{j} - \rho g\mathbf{k}, \tag{6.64}$$

which can be written as

$$\nabla\left(p - \frac{1}{2}\rho\omega^2 x^2 - \frac{1}{2}\rho\omega^2 y^2 + \rho g z\right) = 0. \tag{6.65}$$

Then

$$p - \frac{1}{2}\rho\omega^2 x^2 - \frac{1}{2}\rho\omega^2 y^2 + \rho g z = C. \tag{6.66}$$

The surface of the fluid is determined by $p = p_0$, i.e.,

$$p_0 - \frac{1}{2}\rho\omega^2 x^2 - \frac{1}{2}\rho\omega^2 y^2 + \rho g z = C, \tag{6.67}$$

which is a parabolic surface of revolution. If when $\omega = 0$ the fluid surface is at $z = 0$, then $C = p_0$.

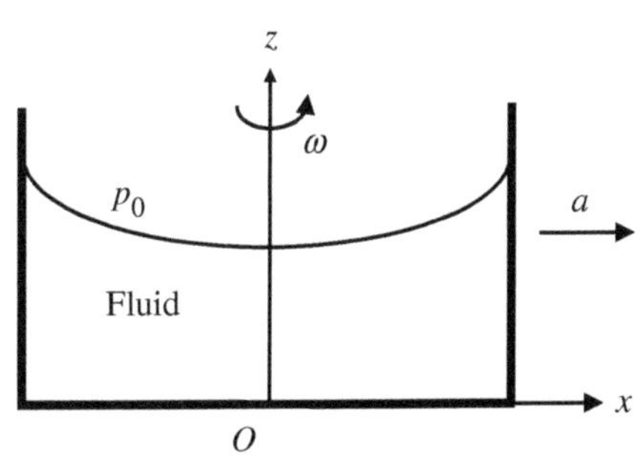

Fig. 6.4. A fluid in a rotating container.

6.9 Couette Flow

Couette flow refers to the flow of a viscous fluid between two parallel plates where one plate is moving relative to the other, creating a shear stress without a pressure gradient (see Fig. 6.5). There is no body force. The fluid is assumed to be incompressible. Consider the possibility of the following velocity and pressure fields which satisfy $v_{k,k} = 0$ (incompressibility):

$$\begin{aligned} v_x &= v_x(y), \quad v_y = v_z = 0, \\ p &= \text{constant}. \end{aligned} \tag{6.68}$$

In this case the Navier–Stokes equation reduces to

$$\nu \frac{\partial^2 v_x}{\partial y^2} = 0, \quad \nu = \frac{\mu}{\rho}, \tag{6.69}$$

whose general solution is

$$v_x = C_1 y + C_2, \tag{6.70}$$

where C_1 and C_2 are undetermined constants. The boundary conditions are

$$v_x(0) = 0, \quad v_x(h) = v_0, \tag{6.71}$$

which determine

$$C_1 = \frac{v_0}{h}, \quad C_2 = 0. \tag{6.72}$$

Hence, the velocity field is

$$v_x = \frac{v_0}{h} y. \tag{6.73}$$

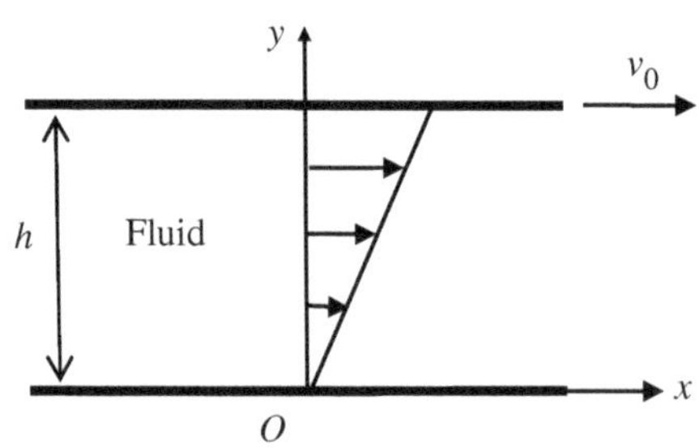

Fig. 6.5. A fluid between a fixed plate and a moving plate.

6.10 Poiseuille Flow

Poiseuille flow is the flow of a viscous fluid through a pipe driven by a constant pressure difference, resulting in a parabolic velocity profile within the pipe. Essentially, Poiseuille flow is driven by a pressure gradient. The similar and simpler situation of a fluid between two plates under a pressure difference is analyzed below (see Fig. 6.6). The fluid is assumed to be incompressible. Consider the possibility of the following velocity and pressure fields which satisfy $v_{k,k} = 0$:

$$\begin{aligned} &v_x = v_x(y), \quad v_y = v_z = 0, \\ &p = p(x), \quad \frac{\partial p}{\partial x} = -\frac{\Delta p}{L} = \text{constant}, \quad \Delta p = p_1 - p_2. \end{aligned} \tag{6.74}$$

The Navier–Stokes equation reduces to

$$-\frac{1}{\rho}\frac{\partial p}{\partial x} + \nu\frac{\partial^2 v_x}{\partial y^2} = 0, \quad \nu = \frac{\mu}{\rho}, \tag{6.75}$$

or

$$\frac{\partial^2 v_x}{\partial y^2} = \frac{1}{\mu}\frac{\partial p}{\partial x} = -\frac{\Delta p}{\mu L} \tag{6.76}$$

The general solution to Eq. (6.76) is

$$v_x = -\frac{\Delta p}{2\mu L}y^2 + C_1 y + C_2, \tag{6.77}$$

where C_1 and C_2 are undetermined constants. The boundary conditions are

$$v_x(0) = 0, \quad v_x(h) = 0, \tag{6.78}$$

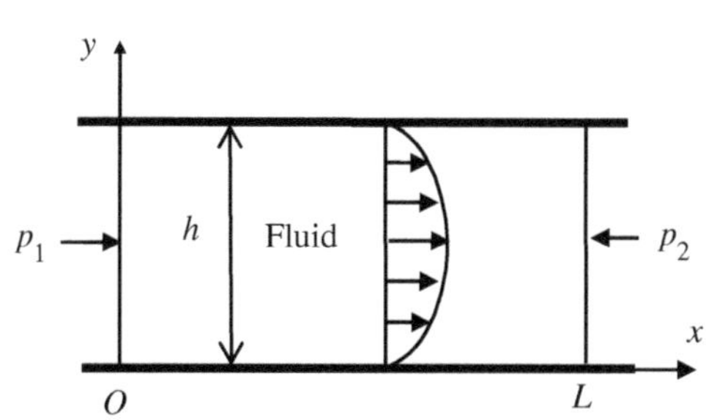

Fig. 6.6. A fluid driven by a pressure gradient.

which determine

$$C_1 = \frac{\Delta p}{2\mu L}h, \quad C_2 = 0. \tag{6.79}$$

Then the velocity field is

$$v_x = \frac{\Delta p}{2\mu L}(h - y)y. \tag{6.80}$$

6.11 Flow Down an Inclined Plane

Consider an incompressible and viscous fluid flowing down an inclined plane under gravity (see Fig. 6.7) [14]. The body force is

$$f_x = g\sin\theta, \quad f_y = -g\cos\theta, \quad f_z = 0. \tag{6.81}$$

Let the trial velocity and pressure fields be

$$\begin{aligned} &v_x = v_x(y), \quad v_y = v_z = 0, \\ &p = p(y), \end{aligned} \tag{6.82}$$

which satisfy $v_{k,k} = 0$. The Navier–Stokes equation reduces to

$$\begin{aligned} &\nu\frac{\partial^2 v_x}{\partial y^2} + g\sin\theta = 0, \\ &-\frac{1}{\rho}\frac{\partial p}{\partial y} - g\cos\theta = 0. \end{aligned} \tag{6.83}$$

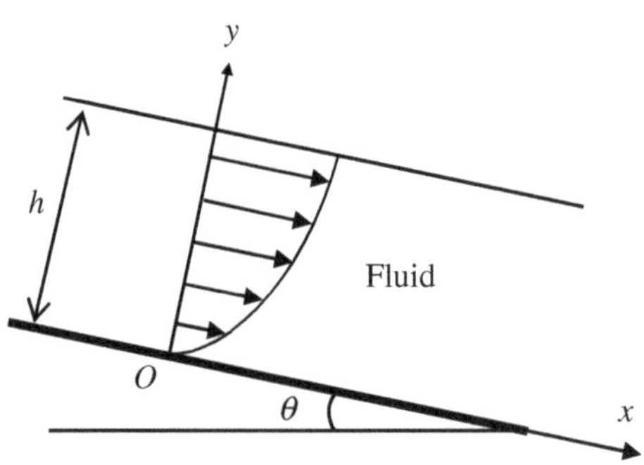

Fig. 6.7. A fluid down a slope.

Integrating Eq. $(6.83)_2$, we obtain the pressure field as

$$p = -\rho g y \cos\theta + C_1, \tag{6.84}$$

where C_1 is an integration constant. From the boundary condition

$$p(h) = p_0, \tag{6.85}$$

we determine

$$C_1 = p_0 + \rho g h \cos\theta. \tag{6.86}$$

Then

$$p = \rho g(h - y)\cos\theta + p_0. \tag{6.87}$$

Equation $(6.83)_1$ can be written as

$$\frac{\partial^2 v_x}{\partial y^2} = -\frac{\rho g}{\mu}\sin\theta. \tag{6.88}$$

Its general, solution is

$$v_x = -\frac{\rho g}{2\mu} y^2 \sin\theta + C_2 y + C_3 \tag{6.89}$$

where C_2 and C_3 are undetermined constants. The relevant shear stress is calculated from

$$\tau_{yx} = \mu v_{x,y} = -\rho g y \sin\theta + \mu C_2. \tag{6.90}$$

From the following boundary conditions:

$$v_x(0) = 0, \quad \tau_{yx}(h) = 0, \tag{6.91}$$

we determine

$$C_2 = \frac{1}{\mu}\rho g h \sin\theta, \quad C_3 = 0. \tag{6.92}$$

Then the velocity and shear stress fields are given by

$$v_x = \frac{\rho g \sin\theta}{2\mu}(2hy - y^2), \tag{6.93}$$

$$\tau_{yx} = \rho g(h - y)\sin\theta. \tag{6.94}$$

6.12 Rayleigh Problem

Rayleigh problem also known as Stokes first problem is a problem of determining the flow created by a sudden movement of an infinitely long plate from rest (see Fig. 6.8). The incompressible and viscous fluid occupies the semi-infinite region of $y > 0$. This is considered as one of the simplest unsteady problems that have an exact solution for the Navier–Stokes equation. Consider the possibility of the following velocity and pressure fields:

$$\begin{aligned} &v_x = v(y,t), \quad v_y = v_z = 0, \\ &p_{,k} = 0. \end{aligned} \tag{6.95}$$

The Navier–Stokes equation reduces to

$$\frac{\partial v}{\partial t} = \nu \frac{\partial^2 v}{\partial y^2}, \quad \nu = \frac{\mu}{\rho}. \tag{6.96}$$

The initial condition is

$$v(y, t = 0) = 0. \tag{6.97}$$

The boundary conditions are

$$\begin{aligned} &v(y = 0, t) = V_0, \\ &v(y \to \infty, t) = 0. \end{aligned} \tag{6.98}$$

Based on a dimensional analysis [15], the trial solution is taken as

$$v = V_0 f(\eta), \quad \eta = \frac{y}{2\sqrt{\nu t}}. \tag{6.99}$$

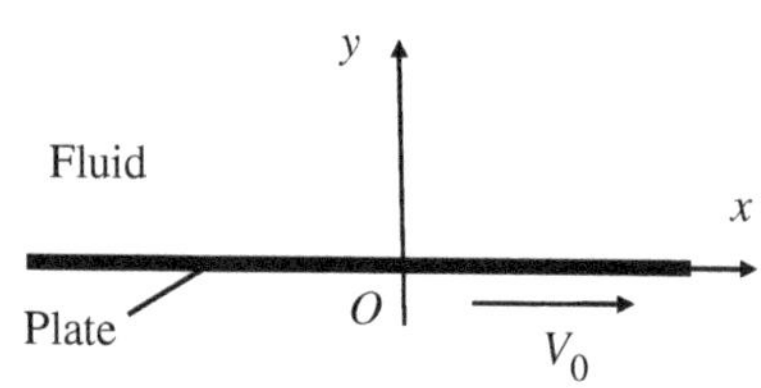

Fig. 6.8. Sudden movement of a plate under a semi-infinite fluid.

Then

$$\frac{\partial}{\partial t} = -\frac{1}{2}\frac{\eta}{t}\frac{d}{d\eta}, \quad \frac{\partial}{\partial y} = \frac{1}{2\sqrt{\nu t}}\frac{d}{d\eta},$$
$$\frac{\partial^2}{\partial y^2} = \frac{1}{4\nu t}\frac{d^2}{d\eta^2}, \tag{6.100}$$

and Eq. (6.96) becomes

$$f'' + 2\eta f' = 0. \tag{6.101}$$

Under Eq. (6.99), the initial and boundary conditions take the following forms:

$$v(y, t = 0) = 0 \quad \Rightarrow \quad f(\eta \to \infty) = 0, \tag{6.102}$$

$$v(y = 0, t) = V_0 \quad \Rightarrow \quad f(0) = 1,$$
$$v(y \to \infty, t) = 0 \quad \Rightarrow \quad f(\eta \to \infty) = 0. \tag{6.103}$$

Integrating Eq. (6.101), we obtain

$$f' = C_1 \exp(-\eta^2), \tag{6.104}$$

$$f = C_1 \int_0^\eta \exp(-\xi^2) d\xi + C_2, \tag{6.105}$$

where C_1 and C_2 are undetermined constants. The application of the initial and boundary conditions leads to

$$f(\eta) = 1 - \mathrm{erf}(\eta), \tag{6.106}$$

where

$$\mathrm{erf}(\eta) = \frac{2}{\sqrt{\pi}} \int_0^\eta \exp(-\xi^2) d\xi \tag{6.107}$$

is the error function with the following properties:

$$\mathrm{erf}(0) = 0, \quad \mathrm{erf}(\infty) = 1. \tag{6.108}$$

Then the velocity field is given by

$$\frac{v_x}{V_0} = 1 - \mathrm{erf}\left(\frac{y}{2\sqrt{2\nu t}}\right). \tag{6.109}$$

6.13 Irrotational Flow of an Ideal and Incompressible Fluid

Irrotational flow, also known as potential flow, is a type of fluid flow where there is no local rotation or vorticity. Consider a nonviscous fluid whose velocity field $\mathbf{v}(\mathbf{x}, t)$ and pressure field $p(\mathbf{x}, t)$ are governed by

$$\dot{\rho} + \rho v_{i,i} = 0, \tag{6.110}$$

$$\frac{\partial \mathbf{v}}{\partial t} + (\mathbf{v} \cdot \nabla)\mathbf{v} = -\frac{1}{\rho}\nabla p. \tag{6.111}$$

For an incompressible fluid, ρ is a constant. We consider irrotational motions. Hence,

$$\nabla \cdot \mathbf{v} = 0, \tag{6.112}$$

$$\nabla \times \mathbf{v} = 0. \tag{6.113}$$

Equation (6.113) allows the introduction of a velocity potential ϕ such that

$$\mathbf{v} = \nabla \phi. \tag{6.114}$$

Then Eq. (6.113) is no longer needed and Eq. (6.112) takes the following form:

$$\nabla^2 \phi = \frac{\partial^2 \phi}{\partial x_1^2} + \frac{\partial^2 \phi}{\partial x_2^2} + \frac{\partial^2 \phi}{\partial x_3^2} = 0. \tag{6.115}$$

With Eq. (6.114), we can integrate Eq. (6.111) to obtain

$$\frac{\partial \phi}{\partial t} + \frac{1}{2}v^2 + \frac{p}{\rho} = f(t), \tag{6.116}$$

where $f(t)$ is an arbitrary function of time and

$$v^2 = \mathbf{v} \cdot \mathbf{v}. \tag{6.117}$$

Once ϕ is obtained from Eq. (6.115), the pressure field is given by Eq. (6.116). On a fixed boundary, the normal derivative of ϕ vanishes. On a free surface, $p = p_0$.

In the special case of a two-dimensional flow without the dependence on z, Eqs. (6.112) and (6.115) reduce to

$$\frac{\partial u}{\partial x} + \frac{\partial v}{\partial y} = 0, \tag{6.118}$$

$$\frac{\partial^2 \phi}{\partial x^2} + \frac{\partial^2 \phi}{\partial y^2} = 0, \tag{6.119}$$

where

$$x = x_1, \quad y = x_2, \quad u = v_1, \quad v = v_2. \tag{6.120}$$

In this case a stream function ψ may be introduced through

$$\begin{aligned} u &= \frac{\partial \psi}{\partial y} = \frac{\partial \phi}{\partial x}, \\ v &= -\frac{\partial \psi}{\partial x} = \frac{\partial \phi}{\partial y}. \end{aligned} \tag{6.121}$$

The relationship between ϕ and ψ in Eq. (6.121) is known as the Cauchy–Riemann condition. $\psi =$ constant defines a curve on which

$$d\psi = \frac{\partial \psi}{\partial x} dx + \frac{\partial \psi}{\partial y} dy = -v dx + u dy = 0, \tag{6.122}$$

or

$$\frac{dx}{u} = \frac{dy}{v}, \tag{6.123}$$

which is a stream line (see Eq. (2.12)). From Eq. (6.121) it can be shown that

$$\frac{\partial^2 \psi}{\partial x^2} + \frac{\partial^2 \psi}{\partial y^2} = 0. \tag{6.124}$$

As an example, consider the steady flow of an ideal and incompressible fluid around a rigid and circular cylinder of radius R in an

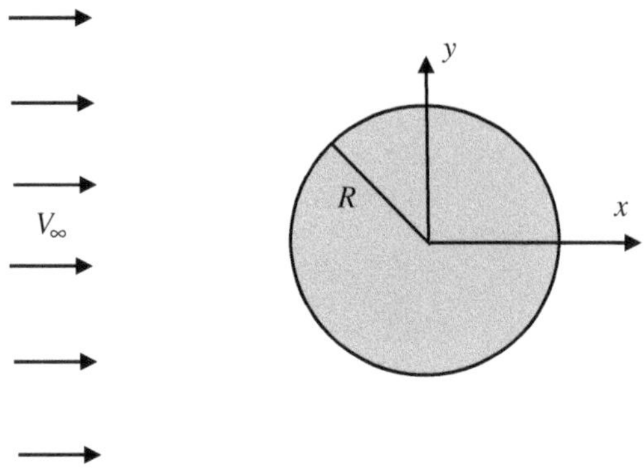

Fig. 6.9. Flow around a circular cylinder.

unbounded domain as shown in Fig. 6.9. In polar coordinates, the boundary-value problem of ψ is

$$\begin{aligned} &\nabla^2\psi = 0, \quad r > R, \\ &\psi = 0, \quad r = R, \\ &\mathbf{v} \to V_\infty \mathbf{i}, \quad r \to \infty. \end{aligned} \tag{6.125}$$

First, we need to determine the far fields in polar coordinates when r is very large. For large r, we have

$$u = \frac{\partial \psi}{\partial y} = V_\infty, \quad v = -\frac{\partial \psi}{\partial x} = 0, \tag{6.126}$$

$$\psi = V_\infty y = V_\infty r \sin\theta. \tag{6.127}$$

Hence, we look for

$$\psi(r, \theta) = \psi(r) \sin\theta. \tag{6.128}$$

ψ is governed by

$$\nabla^2\psi = \frac{\partial^2 \psi}{\partial r^2} + \frac{1}{r}\frac{\partial \psi}{\partial r} + \frac{1}{r^2}\frac{\partial^2 \psi}{\partial \theta^2} = 0, \tag{6.129}$$

which reduces to

$$\nabla^2\psi = \frac{\partial^2 \psi}{\partial r^2} + \frac{1}{r}\frac{\partial \psi}{\partial r} - \frac{1}{r^2}u = 0, \tag{6.130}$$

under Eq. (6.128). Let

$$u = r^n. \tag{6.131}$$

Then

$$\begin{aligned} &n(n-1) + n - 1 = 0, \\ &n^2 = 1, \quad n = \pm 1, \end{aligned} \tag{6.132}$$

$$\psi = \left(C_1 r + \frac{C_2}{r}\right) \sin\theta. \tag{6.133}$$

For the given behavior at large r in Eq. (6.127), we determine that

$$C_1 = V_\infty. \tag{6.134}$$

At $r = R$, the boundary condition in Eq. $(6.125)_2$ requires that

$$\psi = \left(C_1 R + \frac{C_2}{R}\right) \sin\theta = 0, \tag{6.135}$$

which determines

$$C_2 = -C_1 R^2 = -V_\infty R^2. \tag{6.136}$$

Hence,

$$\begin{aligned} \psi &= V_\infty \left(r - \frac{R^2}{r}\right) \sin\theta = V_\infty r \sin\theta - \frac{V_\infty R^2 r \sin\theta}{r^2} \\ &= V_\infty y - \frac{V_\infty R^2 y}{x^2 + y^2}. \end{aligned} \tag{6.137}$$

Problems

6.1. Study the equilibrium of an incompressible fluid in a gravitational field whose potential is given by $V = -A/r$ in spherical coordinates where A is a constant. Determine the pressure field.

6.2. Prove Pascal's law for a confined and incompressible fluid body of any shape in general.

6.3. Prove Archimedes' principle for a partially submerged body.

6.4. Analyze the fluid in Fig. 6.5 under a moving plate and a pressure gradient in the x direction simultaneously.

6.5. Analyze the Poiseuille flow in a circular pipe.

6.6. Determine the steady velocity field of a viscous fluid between two concentric rotating circular cylinders.

6.7. An infinite plate is in contact with a semi-infinite fluid as shown in Fig. 6.8. The fluid is viscous and incompressible. The plate is in time-harmonic motion in the x direction with $V_0 \cos \omega t$. Show that the velocity field in the fluid is given by

$$v_x = V_0 \exp(-ky) \cos(\omega t - ky),$$

$$k = \sqrt{\frac{\omega}{2\nu}}, \quad \nu = \frac{\mu}{\rho}.$$

6.8. For irrotational motions of a barotropic and ideal fluid, show that Euler's equation

$$\frac{\partial \mathbf{v}}{\partial t} + (\mathbf{v} \cdot \nabla)\mathbf{v} = -\frac{1}{\rho}\nabla p + \mathbf{f}$$

admits the following first integral:

$$\frac{\partial \varphi}{\partial t} + \frac{1}{2}v^2 + P + V = f(t),$$

where $f(t)$ is an arbitrary function of time. P and V are defined by Eqs. (6.29) and (6.31).

6.9. For a polytropic fluid with $p = \lambda \rho^\gamma$, show that

$$P = \int \frac{dp}{\rho} = \frac{\gamma}{\gamma - 1}\frac{p}{\rho}.$$

6.10. For an incompressible fluid, show that the Navier–Stokes equation may be written as

$$\rho a_i = \rho f_i - p_{,i} - \mu \varepsilon_{ijk} w_{k,j},$$

$$\mathbf{w} = \frac{1}{2}\nabla \times \mathbf{v}.$$

Appendices

Appendix 1: List of Symbols

$\mathbf{i}, \mathbf{j}, \mathbf{k}(\mathbf{e}_1, \mathbf{e}_2, \mathbf{e}_3)$: Basis vectors of Cartesian coordinates
$\mathbf{i}_1, \mathbf{i}_2, \mathbf{i}_3(\mathbf{I}_1, \mathbf{I}_2, \mathbf{I}_3)$: Basis vectors of Cartesian coordinates
$i, j, k(I, J, K)$: Tensor indices. Range: 1–3
δ_{ij}, δ_{KL}: Kronecker delta
δ_{iK}, δ_{Ki}: Coordinate transformation coefficients
ε_{ijk}, ε_{IJK}: Permutation symbol
p, q: Matrix indices. Range: 1–6
X_K: Reference or material coordinates
y_i: Present or spatial coordinates
$\mathbf{u}$: Displacement vector
F_{iK}: Deformation gradient
J: Jacobian of deformation
C_{KL}, c_{kl}: Deformation tensors
E_{KL}, e_{kl}: Finite strain tensors
I_1, I_2, I_3: Invariants of E_{KL}
ε_{kl}: Linear or infinitesimal strain tensor
ω_{ij}: Linear or infinitesimal rotation tensor
d/dt: Material time derivative
v_i: Velocity vector
a_i: Acceleration vector
d_{ij}: Deformation rate tensor

Ω_{ij}: Spin tensor
ρ^0: Reference mass density
ρ: Present mass density
$\mathbf{t}$: Traction vector
τ_{kl}: Cauchy stress tensor
J_1, J_2, J_3: Invariants of τ_{kl}
τ_{kl}^D: Dissipative or viscous stress
K_{Lj}: First Piola–Kirchhoff stress tensor
P_{KL}, P_{KL}^R, P_{KL}^D: Second Piola–Kirchhoff stress tensor
σ_{kl}, σ_{kl}^R, σ_{kl}^D: Linear or infinitesimal stress tensor
ε: Internal energy per unit mass
U: Internal energy per unit volume
χ: Free energy per unit mass
η: Entropy per unit mass
θ: Absolute temperature
c, c_V, c_p: Specific heat
p: Thermodynamic pressure
p^m: Mechanical pressure
$\mathbf{q}$: Spatial heat flux vector
$\mathbf{Q}$: Material heat flux vector
r: Body heat source per unit mass
κ_{kl}: Heat conduction coefficients
f_j: Body force per unit mass
b_j: Body force per unit volume
E: Young's modulus
ν: Poisson's ratio in elasticity
ν: μ/ρ in fluids
λ, μ: Lamé's elastic constants
λ, μ: Viscosity coefficients in fluids
K: Bulk elastic constant
κ: Bulk viscosity in fluids
c_{ijkl}, c_{pq}: Elastic stiffness
s_{ijkl}, s_{pq}: Elastic compliance
${}_2c_{ABCD}$, ${}_3c_{ABCDEF}$, ${}_4c_{ABCDEFGH}$: Elastic constants
c_1: Speed of longitudinal elastic waves
c_2: Speed of transverse elastic waves

Appendix 2: Cylindrical Coordinates

Cylindrical coordinates (r, θ, z) are defined by

$$x_1 = r\cos\theta, \quad x_2 = r\sin\theta, \quad x_3 = z.$$

The gradient of a scalar field ψ is given by

$$\nabla\psi = \frac{\partial\psi}{\partial r}\mathbf{e}_r + \frac{1}{r}\frac{\partial\psi}{\partial\theta}\mathbf{e}_\theta + \frac{\partial\psi}{\partial z}\mathbf{e}_z.$$

The divergence of a vector field $\mathbf{B}$ is

$$\nabla\cdot\mathbf{B} = \frac{1}{r}(rB_r)_{,r} + \frac{1}{r}B_{\theta,\theta} + B_{z,z}.$$

The Laplace operator (Laplacian) on a scalar field ψ takes the following form:

$$\nabla^2\psi = \frac{1}{r}\frac{\partial}{\partial r}\left(r\frac{\partial\psi}{\partial r}\right) + \frac{1}{r^2}\frac{\partial^2\psi}{\partial\theta^2} + \frac{\partial^2\psi}{\partial z^2}.$$

In cylindrical coordinates, we have the following linear strain-displacement relations:

$$\begin{aligned}
\varepsilon_{rr} = u_{r,r}, \quad \varepsilon_{\theta\theta} &= \frac{1}{r}u_{\theta,\theta} + \frac{u_r}{r}, \quad \varepsilon_{zz} = u_{z,z},\\
2\varepsilon_{r\theta} &= u_{\theta,r} + \frac{1}{r}u_{r,\theta} - \frac{u_\theta}{r},\\
2\varepsilon_{\theta z} &= \frac{1}{r}u_{z,\theta} + u_{\theta,z},\\
2\varepsilon_{zr} &= u_{r,z} + u_{z,r}.
\end{aligned}$$

The equations of motion for linear elasticity are

$$\begin{aligned}
\frac{\partial\sigma_{rr}}{\partial r} + \frac{1}{r}\frac{\partial\sigma_{\theta r}}{\partial\theta} + \frac{\partial\sigma_{zr}}{\partial z} + \frac{\sigma_{rr}-\sigma_{\theta\theta}}{r} + b_r &= \rho^0\ddot{u}_r,\\
\frac{\partial\sigma_{r\theta}}{\partial r} + \frac{1}{r}\frac{\partial\sigma_{\theta\theta}}{\partial\theta} + \frac{\partial\sigma_{z\theta}}{\partial z} + \frac{2}{r}\sigma_{r\theta} + b_\theta &= \rho^0\ddot{u}_\theta,\\
\frac{\partial\sigma_{rz}}{\partial r} + \frac{1}{r}\frac{\partial\sigma_{\theta z}}{\partial\theta} + \frac{\partial\sigma_{zz}}{\partial z} + \frac{1}{r}\sigma_{rz} + b_z &= \rho^0\ddot{u}_z.
\end{aligned}$$

Appendix 3: Spherical Coordinates

The spherical coordinates (r, θ, ϕ) are defined by

$$x_1 = r\sin\theta\cos\phi, \quad x_2 = r\sin\theta\sin\phi, \quad x_3 = r\cos\theta.$$

The gradient of a scalar field is

$$\nabla\psi = -\frac{\partial\psi}{\partial r}\mathbf{e}_r - \frac{1}{r}\frac{\partial\psi}{\partial\theta}\mathbf{e}_\theta - \frac{1}{r\sin\theta}\frac{\partial\psi}{\partial\phi}\mathbf{e}_\phi.$$

The divergence of a vector field $\mathbf{B}$ is

$$\nabla\cdot\mathbf{B} = r^2\frac{\partial}{\partial r}(r^2 B_r) + \frac{1}{r\sin\theta}\frac{\partial}{\partial\theta}(B_\theta\sin\phi) + \frac{1}{r\sin\theta}\frac{\partial}{\partial\phi}B_\phi.$$

In spherical coordinates, we have the following linear strain-displacement relation:

$$\varepsilon_{rr} = \frac{\partial u_r}{\partial r}, \quad \varepsilon_{\theta\theta} = \frac{1}{r}\frac{\partial u_\theta}{\partial\theta} + \frac{u_r}{r},$$

$$\varepsilon_{\phi\phi} = \frac{1}{r\sin\theta}\frac{\partial u_\phi}{\partial\phi} + \frac{u_r}{r} + \frac{u_\theta}{r}\cot\theta,$$

$$2\varepsilon_{r\theta} = \frac{\partial u_\theta}{\partial r} + \frac{1}{r}\frac{\partial u_r}{\partial\theta} - \frac{u_\theta}{r},$$

$$2\varepsilon_{\theta\phi} = \frac{1}{r}\frac{\partial u_\phi}{\partial\theta} + \frac{1}{r\sin\theta}\frac{\partial u_\theta}{\partial\phi} - \frac{u_\phi}{r}\cot\theta,$$

$$2\varepsilon_{\phi r} = \frac{1}{r\sin\theta}\frac{\partial u_r}{\partial\phi} + \frac{\partial u_\phi}{\partial r} - \frac{u_\phi}{r}.$$

The equations of motion for linear elasticity are

$$\frac{\partial\sigma_{rr}}{\partial r} + \frac{1}{r}\frac{\partial\sigma_{\theta r}}{\partial\theta} + \frac{1}{r\sin\theta}\frac{\partial\sigma_{\phi r}}{\partial\phi} + \frac{1}{r}(2\sigma_{rr} - \sigma_{\theta\theta} - \sigma_{\phi\phi} + \sigma_{\theta r}\cot\theta) + b_r = \rho^0\ddot{u}_r,$$

$$\frac{\partial\sigma_{r\theta}}{\partial r} + \frac{1}{r}\frac{\partial\sigma_{\theta\theta}}{\partial\theta} + \frac{1}{r\sin\theta}\frac{\partial\sigma_{\phi\theta}}{\partial\phi} + \frac{1}{r}[3\sigma_{r\theta} + (\sigma_{\theta\theta} - \sigma_{\phi\phi})\cot\theta] + b_\theta = \rho^0\ddot{u}_\theta,$$

$$\frac{\partial \sigma_{r\phi}}{\partial r} + \frac{1}{r}\frac{\partial \sigma_{\theta\phi}}{\partial \theta} + \frac{1}{r\sin\theta}\frac{\partial \sigma_{\phi\phi}}{\partial \phi}$$
$$+ \frac{1}{r}(3\sigma_{r\phi} + 2\sigma_{\theta\phi}\cot\theta) + b_\phi = \rho^0 \ddot{u}_\phi.$$

Appendix 4: Vector Identities

For scalar fields f and g as well as vector fields $\mathbf{a}, \mathbf{b}, \mathbf{c}, \mathbf{d}$ and $\mathbf{G}$, we have

$$\mathbf{a} \times (\mathbf{b} \times \mathbf{c}) = (\mathbf{a} \cdot \mathbf{c})\mathbf{b} - (\mathbf{a} \cdot \mathbf{b})\mathbf{c},$$
$$(\mathbf{a} \times \mathbf{b}) \cdot (\mathbf{c} \times \mathbf{d}) = (\mathbf{a} \cdot \mathbf{c})(\mathbf{b} \cdot \mathbf{d}) - (\mathbf{a} \cdot \mathbf{d})(\mathbf{b} \cdot \mathbf{c}),$$
$$\nabla(fg) = g\nabla f + f\nabla g,$$
$$\nabla \cdot (f\mathbf{a}) = (\nabla f) \cdot \mathbf{a} + f(\nabla \cdot \mathbf{a}),$$
$$\nabla \times (f\mathbf{a}) = (\nabla f) \times \mathbf{a} + f(\nabla \times \mathbf{a}),$$
$$\nabla \cdot (\mathbf{a} \times \mathbf{b}) = (\nabla \times \mathbf{a}) \cdot \mathbf{b} - \mathbf{a} \cdot (\nabla \times \mathbf{b}),$$
$$\nabla \times (\mathbf{a} \times \mathbf{b}) = (\mathbf{b} \cdot \nabla)\mathbf{a} - (\mathbf{a} \cdot \nabla)\mathbf{b} + (\nabla \cdot \mathbf{b})\mathbf{a} - (\nabla \cdot \mathbf{a})\mathbf{b},$$
$$\nabla \times (\nabla \times \mathbf{a}) = \nabla(\nabla \cdot \mathbf{a}) - \nabla^2 \mathbf{a}.$$

For a closed curve C enclosing an area S whose unit normal is $\mathbf{n}$, Stokes' theorem states that

$$\int_C \mathbf{dl} \cdot \mathbf{G} = \int_S (\nabla \times \mathbf{G}) \cdot \mathbf{n} dS.$$

For a closed surface S with an outward unit normal $\mathbf{n}$ enclosing a volume V, the divergence theorem is

$$\int_S \mathbf{n} \cdot \mathbf{G} dS = \int_V \nabla \cdot \mathbf{G} dV.$$

References

[1] X. Du, *Continuum Mechanics*, Tsinghua University Press, Beijing, 1985.
[2] I.S. Sokolnikoff, *Tensor Analysis*, 2nd ed., John Wiley and Sons, New York, 1964.
[3] G.E. Hay, *Vector and Tensor Analysis*, Dover, New York, 1953.
[4] A.J.M. Spencer, *Continuum Mechanics*, Longman, New York, 1980.
[5] A.C. Eringen, *Mechanics of Continua*, 2nd ed., Robert E. Krieger, Huntington, New York, 1980.
[6] I.S. Sokolnikoff, *Mathematical Theory of Elasticity*, 2nd ed., McGraw–Hill, New York, 1956.
[7] W.Z. Chien and K.Y. Ye, *Theory of Elasticity*, Science Press, Beijing, 1956.
[8] K. Feng and Z.C. Shi, *Mathematical Theory of Elastic Structures*, Science Press, Beijing, 1981.
[9] Q.H. Du, S.W. Yu and Z.H. Yao, *Theory of Elasticity*, Science Press, Beijing, China, 1986.
[10] A.C. Eringen and G.A. Maugin, *Electrodynamics of Continua*, Vol. I and Vol. II, Springer–Verlag, New York, 1990.
[11] P. Kelly, *Solid Mechanics Lecture Notes*, The University of Auckland, New Zealand, 2013.
[12] R.D. Mindlin, *An Introduction to the Mathematical Theory of Vibrations of Elastic Plates*, J.S. Yang, ed., World Scientific, Singapore, 2006.
[13] J.S. Yang, *A Concise Course in Elasticity*, Birkhäuser, Switzerland, 2025.
[14] R.C. Batra, *Elements of Continuum Mechanics*, AIAA, Reston, VA, 2006.

[15] R.L. Panton, *Incompressible Flow*, John Wiley and Sons, New York, 1984.
[16] L.D. Landau and E.M. Lifshitz, *Fluid Mechanics*, Pergamon, London, 1959.
[17] L. Meirovitch, *Analytical Methods in Vibrations*, Macmillan, London, 1967.
[18] A.C. Eringen and E.S. Shuhubi, *Elastodynamics, Vol. II. Linear Theory*, Academic Press, New York, 1975.
[19] G.E. Mase, *Schaum's Outline of Theory and Problems of Continuum Mechanics*, McGraw–Hill, New York, 1970.

Index

www.ingramcontent.com/pod-product-compliance
Ingram Content Group UK Ltd.
Pitfield, Milton Keynes, MK11 3LW, UK
UKHW061835290626
6822IPUK00006B/41